T0094151

MATH IN
Drag

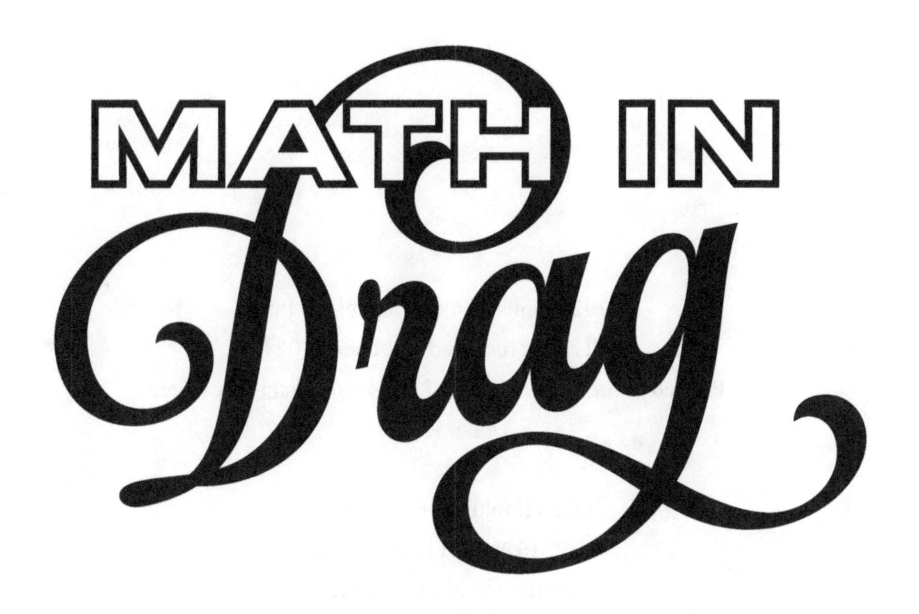

KYNE SANTOS

JOHNS HOPKINS UNIVERSITY PRESS BALTIMORE

Printed in the United States of America on acid-free paper
2 4 6 8 9 7 5 3 1

Johns Hopkins University Press
2715 North Charles Street
Baltimore, Maryland 21218
www.press.jhu.edu

Library of Congress Cataloging-in-Publication Data is available.

A catalog record for this book is available from the British Library.

ISBN 978-1-4214-4874-9 (hardcover)
ISBN 978-1-4214-4875-6 (ebook)

To my parents

CONTENTS

PREFACE

Math is like a drag queen: marvelous, whimsical, at times even controversial, but never boring! On the surface, it may seem confusing or even scary, but when the stage lights turn on, it transforms into an enigma shrouded in wonder and beauty. It's full of twists and turns, reveals and riddles, smoke and mirrors, but encased within it are deep truths about our universe. Math has long been misunderstood and, while technically being accessible to everyone, only a few have seen it up close in all its grandeur. But now's your chance! Step on up because this book is your ticket to the show! The small fee you paid to get your hands on it entitles you to VIP access. I'll be your hostess, emcee, and tour guide as we journey through the magical realm of mathematics.

My name is Kyne, and I'm just your ordinary drag queen who's liked math ever since she was a little boy. My dad was an engineer who practiced my times tables with me and encouraged me to be at the top of my class so I, like him, could experience the joys of optimal engineering. At one point I wanted to become a scientist, then a priest (yes, you read that right), and then an astronomer. I felt like I was capable of being anything . . . except gay.

In the Philippines, where I was born, the word for gay is *bakla*. But more than simply meaning "gay," it also means "a boy who acts like a girl." And honey, was I a big *bakla* growing up! When I got my first Game Boy, fighting and racing games were of zero interest to me; all I wanted was the Disney princess game where I could play as Aurora and Ariel. I used to watch this Filipino girl group on TV called the SexBomb Girls while dancing on my mom's bed, wearing her bra, and using a towel to imitate long, luxurious locks. If I wasn't belting out the group's signature song, Tom Jones's "Sex Bomb," it was Tina Turner's greatest hits on karaoke. I was the star of every family party.

But eventually, I reached an age where all that stopped being cute and started being weird. Soon after moving to Canada, one of my classmates shared that when he told his mom about the boy at school who acted like a girl, she warned him to stay away from me. You know how kids have this innocent voice and, even when they're saying something completely chaotic, it manages to sound playful? That's how he said it. It was the first time I remember thinking there was something wrong with me. I was 7 years old.

Within a few years, people started asking me outright if I were gay. I didn't even know what that word meant! I was too young to feel attracted to anyone, but once I found out that being a "girly boy" was somehow scandalous, I tested the waters with my parents by asking them how they might react if I happened to be gay. They told me that being gay wasn't what God wanted for me and that I'd be better off if I weren't. So I denied it for years and tried to hide any outward signs. But that didn't stop the bullying.

My classmates and I discovered social media right around our 12th birthdays, and it wasn't long before one kid started sending me messages calling me a "homo" and a "faggot" and telling me to kill myself. I never told anyone about it until I wrote it here. I didn't want to disappoint my parents, and I didn't want to disappoint my teachers. Everyone told me I was a good student with a bright future, but I felt as though I was keeping a dark secret. I promised myself that I would grow up, get married, and live a regular life. I used to pray to God, asking why he made me this way, and wish that I could just wake up one day as a girl, so that everything strange about me would suddenly be perfectly normal.

Despite my best efforts, my parents saw right through me and, when I was 14 years old, they sat me down and asked me to come out of the closet. Honey, the jig was up for this *bakla*! I was completely horrified, but with that moment of horror came a sigh of relief. My parents had freed me from the heavy burden I had been carrying on my shoulders. Our conversation went on for hours, and they asked questions like, "How long have you known?" and, "Why didn't you tell us sooner?"

The truth was that, as a child, I felt like my parents were suppressing my true *bakla* self, and it hurt. My dad in particular would try to stop me from wearing bright colors or walking around with a limp wrist. They wanted to protect me from bullies, but I was stronger than they gave me credit for.

All I ever wanted was their support, so once they knew the truth, I didn't really care about what anyone else thought of me. I started watching makeup tutorials online and teaching myself how to apply foundation and eyeshadow. I googled questions like, "Is it okay for boys to wear makeup?" and discovered other boys like me on YouTube who wanted to play with cosmetics.

I soon wore lipstick to school and light concealer over my pimples, and I started my own YouTube channel to document my makeup journey. I had no clue what I was doing, but I knew I loved the feeling of putting on cosmetics. Recording my efforts on camera for the amusement of others was a good excuse to practice. Soon tinted moisturizer and powder evolved into neon pink *ARTPOP*-inspired eyeshadow.

Makeup was my creative outlet for self-expression—the superhero costume that gave me the confidence to walk with my chin up. At school dances, I wore leather pants and a pair of high heels I borrowed from a friend. I even glued rhinestones to my face when glitter and false lashes weren't providing enough drama for me. I was totally comfortable being gawked at by everyone and fitting in with no one. I made my face a canvas for art, and the school hallways were my runway. I did an entire lip-sync performance to Lady Gaga's "Applause" at my (Catholic) high school's Christmas concert (see the fierce photo on page ix). My parents' reaction was along the lines of, "We get that you're gay, but do you have to be *that* gay?" What can I say? When I came out of the closet, I had to make up for lost time!

My performance ended with me jumping off the stage and landing in the splits on the gymnasium floor. Let me tell you, everybody in that audience gagged. All the girls were thrilled, and the triumph was memorialized in my yearbook.

Not everyone was a fan of my antics, but by then the bullies were in my rearview mirror. I had realized that I was an undiscovered supermodel; everyone else just hadn't caught on yet. After feeling like an outcast and an underdog for so many years, I resolved to ace every course. I wanted to be fierce in the classroom, too! I had no shame in being what kids would call a "try-hard." My newfound confidence led me to a circle of friends just as studious and ambitious as I was, and we encouraged each other to get As on every exam and assignment. I suppose I wanted to prove to myself that I could still excel in school despite being out, flouncy, and flamboyant.

In English class, I presented a book report on Goethe's *Faust* wearing a long gray lace front wig, beard, and prosthetic makeup, along with white contact lenses in my eyes and a solar system diorama on my head to represent the aged Faust's enlightenment in heaven. That was quite the bus ride to school!

For our assignment on Shakespeare's *Macbeth*, my friends and I made a music video in which we sang and danced to Lady Gaga's "Bad Romance." I know what you're thinking but, no, I didn't play Lady Macbeth. Sadly, my friend won that role, while I played Macbeth himself. However, I did talk my way into being filmed taking a milk bath while my costar dropped in red food coloring, symbolizing Macbeth's descent into violence. My teacher, Mrs. Cowan, was living for it!

But my favorite class by far was math. While it was much less theatrical, I found the quiet solitude of algorithmically solving equations quite satisfying. I liked how I could ace the tests and assignments by simply getting all the questions right, unlike in English class, where my teacher could deduct marks for what, I felt, were subjective opinions. I wasn't sure what all those equations and theorems were for, and I didn't know how my dad used them in his job, but I knew I was good at it, and that made him happy.

My teachers pulled me aside and told me about math competitions I could enter to challenge myself further, and it was these contests that really kindled my love affair with mathematics. They were definitely harder than the tests I took during class, and they required more creative thinking. I discovered that math required artistry, and I began to see it not just as a subject in school but as a fixture of nature, a real mystery to be solved. Once I recognized the same element of creativity in math that was so compelling to me in my adventures with makeup, there was no holding me back. In a competition during my junior year, I earned one of the top three scores in the world and, as a result, was invited to attend a weeklong math camp at the University of Waterloo. Rest assured, I was the most glamorous mathie there.

That week introduced me to the world of higher-level math—the kind that isn't all about numbers and rules and getting the right answer but about patterns, proofs, and innovative thinking. Instead of memorizing formulas or performing functions like a calculator, I had to push the limits of my imagination. It was as challenging as it was rewarding. That was when I decided I wanted to pursue a math degree. I wasn't sure exactly what I wanted to *do* with a math degree, but everybody told me that it would help me land a good job. The proof of this came quickly; by my second year of university, my friends were building apps, starting their own businesses, and interning for big banks. I was ready to begin building a professional path as well.

So I became a drag queen.

The idea was that I could have some fun first, and then later figure out how to make money. One day, strictly as a joke, I started posting videos on TikTok of me explaining math riddles while dressed in drag. Until then, I'd never entertained the thought that there was any direct connection between the world of drag and the world of math, but I soon started receiving messages from people all over the world telling me that I had helped them finally relate to a subject they hadn't realized was so interesting.

I grew up with the unfortunate stereotype that math is a boys' club, particularly a straight boys' club, and that any woman or queer person who wanted to be part of it must adapt to straight-boy sensibilities. But the more I learn about math, the more I see the connection with drag—math is art, chaos, elegance, freedom, creativity, and abstraction. It's so much more than just scrambling numbers and symbols around to get the right answer. Sometimes there's no right answer at all.

I wrote this book to introduce you to this mathematical multiverse. Instead of making you memorize formulas or perform calculations, I'll tell you the formulas' backstories, and how they're used in the world around us. It's worth noting that some of these origin stories are likely mythical, or at least heavily embellished over thousands of years of oral retellings. I include them anyway because I think they infuse life into the math we discuss. Just take the histories with a couple grains of salt . . . followed by a shot of tequila and a wedge of lime!

This is absolutely *not* a textbook but rather a simple drag queen's guide to the world of math. You don't need to be a mathematician to read this book, just as you don't need a five-octave vocal range to listen to Mariah Carey and don't need to be a queen to have fun at a drag show. All you need is an imagination and a generous dash of curiosity.

In a rough sense, the topics I cover in this book can be placed in one of two categories: familiar math and unfamiliar math. I'll recontextualize math you may already be familiar with in more fabulous and revealing ways, and I'll introduce to you some exciting, unfamiliar math that perhaps has yet to make its way into classrooms. The chapters are organized so that we can bounce back and forth between those two categories. We'll spin through some subjects you may have encountered in high school, such as probability, geometry, and statistics, as well as topics covered in undergraduate math courses, such as the nature of infinity and complex numbers. Be prepared to linger in neighborhoods where math intersects with art, life, and drag.

We begin chapter 1 in the realm of infinity because it's a familiar concept in some ways, but in other ways, it's clandestine knowledge. While there, we'll meet busloads of queens and monkeys older than time. Then, in chapter 2, we'll go full paparazzi on what I call "celebrity numbers"—the rebels, punks, and stars of the number world. In chapter 3, we leave the abstract to share a cake with some legendary pop stars and calculate the chances of winning *RuPaul's Drag Race* (yes, we're going there). In chapters 4 and 5, we'll dive down the rabbit hole of randomness, try to win a Pokémon battle, and examine the statistical anomaly that is queerness. We just might figure out how to harness the power of exponential growth to change the world by the end of chapter 6. Then, in chapter 7, our whirlwind tour ventures into the uncharted territory of illegal math to observe the ways that great math, just like great drag, breaks all rules and expectations when let loose into the wild. You might not think of geometry as a path to the secrets of the universe, or an allegory for queer liberation, but don't worry, I'll show you the way in chapter 8. We'll sit down for a little existential chat in chapter 9 about whether numbers are even real, whether any solution can be objectively true or false, and what the ballroom scene has to say about it. Before we say goodbye, my final curtain call in chapter 10 is meant to leave you thoughtful, and maybe even in the mood for applause.

When I set out to write this book, my primary goal was to convert as many people as possible into loud and proud math lovers, especially people who might have been previously scared away because they don't fit society's mold for someone who's "good at math." As I began writing, though, a second goal materialized, one born out of my growing concern about a reemerging culture of fear and ignorance toward queer people, a culture we'd hoped had been eradicated. With some of the largest governments in the world taking legislative steps to push queer people out of all sectors of public life, I want this book to function as my testament to our humanity.

Let me get some terminology out of the way while we're all here. I am writing as a gay man who performs in drag, and my pronouns are he/him when out of drag and she/her when I'm dressed in my full regalia. But you can call me he, she, they, it, or whatever you please. I am queer, meaning that I take Celine Dion extremely seriously. But to get specific, "queer" means that I belong to the LGBT community, a group of people united by a shared culture that is detached from traditional ideas of gender, sex, and love. This includes lesbians, gays, bisexuals, transgender people, intersex people, two-spirit people, and all other sexual minorities who identify with the community. I will frequently use "queer" as an umbrella term for all members of the community, and I will sometimes use "gay" to mean the same. Some members of the community reject the use of one or the other or even both terms as general identifiers. Please note that I don't speak on behalf of the entire community, and I can only write through my own limited perspective. But as for me, I am queer and gay, meaning that I'm a proud member of this community, and I'm also "odd, peculiar, and happy."

Now let's kick off this tour of the wondrous world of mathematics! Regardless of how you feel about math, or about gay Filipino drag queens, I invite you to read this book with an open heart and an open mind. I hope that you learn something new, whether it be about the world or about yourself.

MATH IN
Drag

CHAPTER 1

Infinite Possibilities

Being a world-famous, microcelebrity drag queen comes with perks. You get to meet adoring fans, who roll out the red carpet when you perform in their ultra-glam school cafeterias or the local public library's parking lot. I may have millions of followers online, but I frequently dazzle live audiences that number in the tens. I once appeared at a small-town Pride celebration in the paved area behind a senior citizens' recreation center, where half of the 20 people in the audience were event organizers and volunteers. All six performers shared a dressing room—a camping tent tucked behind a barrel of hay—which kept getting blown away by chilly gusts of wind. Still, whether you're up there in front of 20 people or 2,000, you must always give it your all!

To me, a full theater of 200 people feels like a huge crowd, but I've attended concerts with audiences of more than 10,000 in stadiums so big that everyone looked like ants. It's hard to imagine what a million people together in one place would look like: a crowd 100 times larger than a 10,000-seat stadium!

On the internet, it's easy for videos to be viewed by hundreds of millions of people all across the world, but this is still only a small fraction of the 8 billion people living on our planet. Of course, larger numbers than billions exist, and writing them down would be even harder than envisioning them if we hadn't developed a handy notation system to save us from being overwhelmed by zeros. Numbers bigger than a billion can be expressed based on the number of digits they have. A trillion (1,000,000,000,000) has 13 digits, or a 1 followed by 12 zeros. Another way to write this is 10^{12}. The number of blades of grass in the world is approximately 10^{18}, or a 1 followed by 18 zeros (see the appendix for a fun calculation). The number 10^{100} (a 1 followed by 100 zeros) is called a "googol" (pronounced just like the famous search engine) and looks like:

10,000,000,000,000,000,000,000,000,000,000,
000,000,000,000,000,000,000,000,000,000,000,
000,000,000,000,000,000,000,000,000,000,000

The term was introduced by the American mathematician Edward Kasner, the first Jewish faculty member in the sciences at Columbia University. If "googol" sounds playful for a math term, that may be because Kasner's 9-year-old nephew, Milton, dreamed it up one day while on a walk in the park with his uncle. To conceptualize the immensity of a googol, imagine adding a pile of electrons (particles even smaller than an atom) onto a scale one at a time until the pile becomes as heavy as the mass of the entire visible universe. It would take about 10^{90} electrons, which is *still less than a billionth of a googol.*

There are numbers too large to even fit on this page, like Kasner's googolplex, which is 10 to the power of a googol, $10^{10^{100}}$. That's a 1 followed by a googol zeros. There isn't enough material available in the entire universe to even write this number down, so how could we possibly try to imagine a googolplex of anything? Yet I'm sure you can still think of even bigger numbers, like a googolplex plus one, or a googolplex to the power of a googolplex. You could keep counting as high as you like, all day, every day, for the rest of your life; your only limitation is your mortality. But all these numbers pale in comparison to *infinity*.

Infinity is enchanting, mystifying, and strange, and yet even young children are able to grasp some notion of it, as in the idea of eternal life or Willy Wonka's Everlasting Gobstopper. One way we often think about infinity is through the idea of a never-ending process, like the digits of pi (π) repeating endlessly:

$$\pi = 3.14159265359\ldots$$

Technically, every number can be written with infinitely many digits.

$$5 = 5.0000000000\ldots$$

$$\frac{3}{5} = 0.600000000000\ldots$$

$$\frac{1}{3} = 0.33333333333\ldots$$

$$\frac{2}{7} = 0.285714285714\ldots$$

These numbers all have infinitely many *repeating* digits, whether it be a repeating 0, a repeating 3, or a repeating pattern of 285714, over and over again. But there are other numbers with infinite digits, which, like the celebrated π, seem to parade forward forever randomly, with no apparent pattern or repetition at all. These numbers are called *irrational*.

An urban legend claims that somewhere within all the infinite digits of π are your phone number, your birthday, your credit card number, and every other string of numbers you can think of. Yet as astounding as that would be, it wouldn't make π unique. It would still only be one of a set of what mathematicians call "normal numbers." The key element of normal numbers is that they feature every digit from 0 to 9 with equal frequency. While π is likely a normal number, mathematicians have yet to prove that. There are infinitely many normal numbers, but we only know of a few, such as the Champernowne constant:

$$0.12345678910111213141516171819202122232425\ldots$$

The Champernowne constant starts at 1, then ticks along to 2, then 3, proceeding all the way along the number line forever. A number like the Champernowne constant, or π (3.141592653589...) would be considered "normal" if the 9s appear just as often as the 1s do, and the 5s appear just as often as the 4s do, and in general every string of numbers is equally likely to appear as any other. Infinity is an awe-inspiring power all on its own, but when randomness and infinity are combined, anything is possible. Just ask Paul the Octopus!

Animal Oracles and Infinite Monkeys

In 2010, an octopus in a German aquarium by the name of Paul gained international fame by correctly predicting the winners of 8 FIFA World Cup games in a row. To obtain his incredible insights, his captors placed two boxes of food in his tank, with each box bearing a flag of one of the teams playing. The box of food that Paul opened first was taken to be Paul's prediction of who would win that match. Correctly predicting the winner of a single game is as likely as winning a coin flip—there's a 50% chance of getting it right. However, correctly predicting multiple games in a row is harder.

As Paul's streak of correct predictions grew longer, so did the media attention. Eventually he grew into a German celebrity. When Paul predicted that the German team would lose to Spain in the semifinals, the Germans called for him to be eaten, and the Spanish prime minister joked about sending Paul a bodyguard.

But how special is this octopus, really? Correctly predicting the outcome of 8 games is no easy feat—it's as likely as winning 8 coin flips in a row, a probability of only 1 in 256. It was this low probability that led many fans to believe that Paul was psychic, or at least extremely lucky.

The thing is, there are plenty of zookeepers and pet owners around the globe who try to predict the outcome of sports games via the use of animal oracles, like Leon the Porcupine, Carly the Chicken, Mani the Parakeet, and Petty the Pygmy Hippopotamus, to name a few. They predict all sorts of outcomes, as you can imagine, but it's the ones who happen to get it right that make headlines, like the tentacled prophet, Paul.

Now imagine a room filled with infinitely many animal oracles tasked with predicting the future by randomly selecting boxes of food. They would predict infinitely many futures to choose from, and one of them is bound to be right. Or we could replace them all with only a single animal oracle that's given an infinite length of time to make endless predictions, say, an immortal monkey. Every day, the monkey presses random keys on a keyboard, for all eternity. Eventually, those random clicks would produce the entire Bible word for word, all the sonnets of Shakespeare, the lyrics of every song in Madonna's entire discography, and the text of this very book you're reading.

To show you how this idea—which mathematicians call the "infinite monkey theorem"—works, let's consider a simplified keyboard consisting of 26 capital letters, a spacebar, a comma, and a period, for a total of 29 characters. In our experiment, we'd like to know the possibility of our monkey typing VOGUE, which contains only 5 characters. The probability that the monkey will randomly hit the letter V is 1 in 29, and the probability that he will type the letter O is also 1 in 29. To see the probability of him typing these characters immediately after one another, we need to multiply the probabilities together; so the probability of typing the 2-character string VO is $(1/29) \times (1/29) = 1/841$ or about 0.12%. If we want him to type the entire string VOGUE, we need to multiply 1/29 by itself for a total of 5 times, giving us a probability of:

$$\left(\frac{1}{29}\right)^5 = 0.0000049\%$$

While this is a very small probability, it's not zero! Of course, it's much more likely that the first 5 keys the monkey hits will look like complete gibberish to us, such as JVBWO. Mathematically, the chances of the monkey *not* typing our target phrase is 99.9999951%. What if, after hitting the first 5 characters, the monkey gives it another try and types 5 more characters? There's another 99.9999951% chance he gets it wrong a second time. But what if the monkey tries a million times? Suppose that our immortal monkey can type 1 character per second, never once stopping to sleep, eat, or (justifiably) attempt to escape. In that case, each attempt at 5 characters takes the monkey 5 seconds. His 1 million attempts at typing VOGUE would take about 58 days. After 58 days, the chances that the monkey will have *missed* the target phrase each time is equivalent to

$$(0.999999951)^{1,000,000} = 0.952, \text{ or } 95\%$$

What if we wait 580 days, and give him 10 million tries? Well, then the probability that he *didn't* type VOGUE would fall from 95% to 61%. If we are willing to wait even longer, say 16 years (which is 100 million tries), the probability of missing VOGUE each and every time falls to less than 1% (0.74%). As the monkey goes on typing forever, cursing the questionable life choices that led him to this fate, the probability of missing VOGUE gets smaller and smaller, meaning that given enough time, typing that 5-letter string effectively becomes inevitable.

It may come sooner than "forever" if you allow VOGUE to have been typed across two different strings of 5 letters. Suppose that the monkey's first attempt at typing 5 letters is

JDKVO, and his second attempt is GUESP. We might categorize that as two failed attempts, but when put together, VOGUE appears in the middle:

JDKVO GUESP

Searching for VOGUE in between two separate strings of 5 characters will require some harder math though, so let's stick with our simpler model. After all, the monkey has all the time in the world, so why rush? If it takes 16 years for the monkey to have a 99.26% chance of typing VOGUE, how about LA ISLA BONITA? That song title has 14 characters, and to ensure a 97% chance of having gotten it right at least once will take over 400 trillion years. (See the breakdown in the appendix.)

We can ask the monkey to type out every single lyric Madonna has ever sung, and we can be almost certain that the monkey will type them eventually, even if we have to wait a googolplex number of years or longer. After enough typing, the monkey will have produced a transcript of everything you have ever said from the day you started speaking up until your last word. He will also have typed out every profound thought that's made its way through your mind because all of these are simply different permutations of the alphabet, plus some spaces and punctuation marks. If only we had infinite time on our hands, each of us would be capable of making masterpieces.

Infinity is so long that the monkey doesn't even have to type one key per second. The keyboard hits could be spaced further apart, and every possible text would still get typed eventually. The monkey could leverage his infinite seniority to choose to only work on Tuesdays and it wouldn't be a problem. An infinite number of Tuesdays is just as infinite as an infinite number of weekdays. For all we care, the monkey could hit only a single key on every February 29, spending the rest of his time on vacation in Bali, and we would still reach every possible goal text. A presumably much less cranky monkey could still supply us with an eternity's worth of labor by only pressing one key every four years.

I have to admit I don't like the idea of unpaid labor, or holding animals in captivity, much less for an eternity. So I think the least we can do is compensate the monkey with a $100 bill for every key he presses. I don't want him to have to work faster to make more money, though, so I'll offer to pay him up front with an infinite supply of $100 bills to press infinitely many keys. But suppose that one of my monkey boss colleagues is a lot more callous and suggests lowering that wage to only a $1 bill per key, on the argument that both wages equate to an equally infinite sum of money. Which option should the monkey take—the infinite supply of $1 bills or the infinite supply of $100 bills? One of these options surely looks like more money, but how do you compare infinite amounts? Is 100 times infinity just infinity itself, or are there different sizes of infinity?

The Cardinality of Queens

Infinity may only be an abstract idea in our everyday language, but mathematicians have spent a great deal of time (although only a finite amount of it) working out exactly how to contend with infinity as a mathematical concept. To a mathematician, an infinite supply of $1 bills is just as much money as an infinite supply of $100 bills, even though they seem different.

This perception is based on the work of nineteenth-century Russian mathematician Georg Cantor. Cantor was the first to start treating infinity as a formal mathematical object that had its own set of logical rules. To do so, he had to create a whole new branch of mathematics called set theory.

A *set* is any collection of definite and separate objects or elements that you can imagine. We can name them one by one, as in the sets {1, 2, 3, 4}, {cat, dog, bird, fleas}, or {lashes, heels, lipstick, perfume}, or we can simply describe them without naming each member individually. My playlist of favorite songs is a set, and so is the set of all odd numbers between 1,000 and 2,000. Sets can be empty, and sets can contain other sets. Sets can even contain themselves as members.

Cantor's great innovation was in devising a method for counting sets. He introduced the notion of *cardinality*, which is defined by the number of elements within a given set. My Spotify playlist "Quintessentially Kyne" has 79 songs (all of which I've listed in the back of the book for your enjoyment), so it's a set with a cardinality of 79. The empty set has a cardinality of 0, but a set which contains the empty set has a cardinality of 1; it still contains a set, after all, even if it's an empty one. The set {2, 3, 4, 5} has a cardinality of 4. Cantor proposed that two sets have the same cardinality if it's possible to pair up every element of one set with an element from the other.

Suppose you have a set of wigs in a box and also a set of queens who you would like to use as wig models, and you want to be certain that you have as many wigs as you have queens. In other words, you want to compare their cardinalities. Of course, you could just count each set individually, but another method is to have each queen simply put on a wig. You will then be left with one of three possibilities:

1. If there are any wigs left over, then there must have been more wigs than queens, and the cardinality of wigs was greater than the cardinality of queens.
2. If there are no wigs left over but there is at least one unhappy, wigless queen in the room, then you'll know that the cardinality of queens was greater than the cardinality of the wigs since there weren't enough wigs to go around.
3. If every queen is bewigged and there are no wigs left over, then there were equally as many queens as there were wigs, and their cardinalities were the same.

Imagine you had a troupe of 100 queens and a room filled with 100 wigs. If a new queen entered the scene without a wig, you'd immediately know that there would be no way to accommodate her, since we would have one too many queens—or one too few wigs! Our unfortunate newcomer would have to go wigless for the time being.

But sets don't have to contain a finite number of wigs or a finite number of queens. They can be infinitely large, like the set of all odd numbers greater than 573, or the set of all points on a circle! Suppose that you work for Kyne's Grand Wig Emporium, a wig store which has an infinite number of wigs of all different styles, which are uniquely numbered from 1, 2, 3, 4, through to infinity. To model your wigs, you find a club full of infinitely many queens. As long as each queen has a wig to wear and each wig has a queen to wear it, according to Cantor's rule of matching (which mathematicians call a bijection), these sets have the same cardinality.

Here's where things get fun. Imagine that now a new queen enters the mix. Can we find her a wig even though all the wigs are currently taken? It seems like it shouldn't be possible, but because of infinity, it is! With some clever shuffling of wigs, we can accommodate the newbie. Sure, there's bound to be some fighting over who gets to wear what—someone will say, "I don't do blonds!" or "This style makes me look hideous!" It could degenerate into wig wars and someone could get hurt.

To lessen the likelihood of bloodshed, we can ask the queen who's currently wearing wig 1 to switch to wig 2, and ask the queen wearing wig 2 to switch to wig 3, and for the one wearing wig 3 to switch to wig 4, and so on, asking each queen to swap to the next wig down the line. Since there are infinitely many queens, there is no "last queen" who gets snubbed. Every queen will have a wig to wear, and wig 1 will be free for the new queen to try on. Every queen will be wearing a wig, and every wig will be worn by a queen, so both sets have the same cardinality, even after adding a new queen to the mix.

Now suppose that there's been a she-mergency, and our wig manufacturer notifies us that some of the wigs are defective. Due to a computer error, every odd-numbered wig has been contaminated with itch powder and needs to be recalled. Since there are infinitely many odd numbers {1, 3, 5, 7, 9, 11, ...}, we have to give up infinitely many wigs. It would seem like this is a major problem because half of our queens would go wigless! But it's no problem at all; we can simply shuffle things around again to give every queen a wig. The queen who had to give up wig 1 can switch to wig 2. The queen who was wearing wig 2 can switch to wig 4. The queen who was wearing wig 3 can switch to wig 6, and so on.

$$\begin{gathered}\text{Previous wig number} \rightarrow \text{New wig number}\\ 1 \rightarrow 2\\ 2 \rightarrow 4\\ 3 \rightarrow 6\\ 4 \rightarrow 8\\ 5 \rightarrow 10\\ \ldots\\ \mathrm{N} \rightarrow 2\mathrm{N}\end{gathered}$$

In general, a queen who was wearing a wig numbered N can switch to the wig numbered 2 times N. If everyone agrees to this switch, we will have successfully eliminated all odd-numbered wigs, while still ensuring that every queen has a wig on her head and every wig has a queen to wear it. As before, both sets have the same cardinality, even after we slashed one of those sets in half!

What we've shown here is that the cardinality of the set of *natural numbers* {1, 2, 3, 4, 5, ...} is the same as the cardinality of the set of positive even numbers {2, 4, 6, 8, 10, 12, ...}!

Natural numbers are the positive whole numbers: 1, 2, 3, 4, and so on. You can think of them as the counting numbers, or the numbers that feel the most "natural," whatever that means. Essentially, our thought experiment demonstrates that there are as many even numbers as there are natural numbers, even though one of these is a subset of the other! This can only happen because of the strange mechanics of infinity.

Now imagine a different, more serious problem. Due to a scheduling error, an infinite number of buses arrive at the Wig Emporium, and each bus is carrying an infinite number of queens on it, and all of these queens would like a wig of their own. There are as many buses as there are wigs, so it would seem as though we only have enough wigs to give one to each bus. Just on bus 1 alone, there are enough queens to use up the entire supply! How could we possibly accommodate every single queen on every single bus with their own unique wig? Amazingly, there's still a way to create a one-to-one matching here! We can number the buses, as well as each seat number, starting from 1. Each queen will combine their bus number with their seat number to get a unique identification (ID) number for the wig they will eventually take for themselves, and that unique wig ID looks like this:

[first digit of bus number] [first digit of seat number]
[second digit of bus number]
[second digit of seat number] etc.

For example, the queen sitting in bus 892 on seat **427** will take wig 8**4**9**2**2**7**. The queen sitting in bus 67192 on seat **10508** will take wig 6**1**7**0**1**5**9**0**2**8**.

However, there's a small problem here, because there won't be anyone to wear the single-digit wigs, like wig 1. To fix that, we can relabel the buses to start with bus 0, then bus 1, bus 2, and so on. That way, the queen sitting on bus 0 in seat **1** will wear wig 0**1** (or simply wig 1), and the queen sitting on bus 0 in seat **10** will wear 0**1**0**0** (or wig 100).

This method ensures that every queen on every bus gets a different wig, and that every single wig gets used because we can trace each wig number to a different queen with their exact bus number and seat number. Every queen will be assigned to a wig, and every wig is assigned a queen. The cardinalities of both sets are the same! This means that an infinite supply of queens has just as many queens as there are in an infinite supply of buses each filled with infinitely many seats! Infinity . . . times infinity . . . is still infinity!

Transfinite Numbers

Mathematicians have given this size of infinity a name: *aleph-0*, pronounced "aleph-null" or sometimes "aleph-naught" or "aleph-zero," and shown as $\aleph_0$. The number $\aleph_0$ doesn't fit within our typical idea of how a number should behave. It doesn't exist anywhere on the number line. It is a new kind of number, one that defies our traditional perceptions of what numbers can be. It is known as a *transfinite* number. Just like our beautifully bewigged drag performers, transfinite numbers have their own rules, which is why we can write the following sentences, which would not make sense for any other numbers:

$$\aleph_0 + 1 = \aleph_0$$
$$\aleph_0 + \aleph_0 = \aleph_0$$
$$\aleph_0 \times 100 = \aleph_0$$
$$\aleph_0 \times \aleph_0 = \aleph_0$$

$\aleph_0$ is larger than any natural number. It is, in fact, the number of natural numbers, the number of even numbers, and also the number of odd numbers. It is the number of seconds in an eternity, and the number of years in an eternity. That's why our immortal monkey can get away with pressing our keyboard only once every thousand Tuesdays, and still eventually write every piece of finite text possible. $\aleph_0$ is also the number of dollars in an infinite supply of $1 bills, and also the number of dollars in an infinite supply of $100 bills. It is the number of queens in an infinitely large bus, but also the number of queens in infinitely many buses, each containing infinitely many seats.

The number $\aleph_0$ is incomprehensibly large. And yet, it is considered to be the very smallest infinity. There are other infinite sets which are even larger than the set of natural numbers, such as the *power set* of the natural numbers. So how does a power set work?

If S is any set, then the power set of S is the set of all subsets of S. I know this may sound wild to some of you, so let me explain with an example. Let S be the set {Adore, Bianca, Courtney}. There are 8 possible subsets we can make from this set:

- {} (the empty set)
- {Adore}
- {Bianca}
- {Courtney}
- {Adore, Bianca}
- {Adore, Courtney}
- {Bianca, Courtney}
- {Adore, Bianca, Courtney}

The original set {Adore, Bianca, Courtney} contains 3 members, whereas the power set contains 8 members, which is equivalent to 2^3. A subset can contain all the members of the original set, or none of them, or only some of them. Importantly, we don't count reorderings as a distinct subset, so for instance {Adore, Bianca} is identical to {Bianca, Adore}.

You can think of subsets as cliques, circles, or "houses" that exist within the larger set. I'm borrowing the concept of houses here from ballroom culture, which originated with the drag balls of the 1800s among Black queer people in the United States and later crossed over into drag culture. Invented by and primarily attended by Black gay men who were formerly enslaved or the children of enslaved people, today houses function as chosen families for queer people, especially people of color, who may have been ostracized from their traditional families. As the legendary queen Pepper LaBeija put it in the famous 1990 documentary *Paris Is Burning*, "A house is a family for those who don't have a family."[1] In the contemporary ballroom scene, houses are sometimes named after famous fashion houses, like the House of Revlon or the House of Balenciaga.

For the purposes of our mathematical analogy, we'll say that a "house" can be any subset of our infinite set of queens. A house may be finite, empty, or infinitely large, and each queen is allowed to belong to multiple houses at the same time. One possible house might be {queen 1, queen 3, queen 5}, or {queen 1,892, queen 567}. To write down all the houses, we can use an infinitely large spreadsheet, where the queens' numbers form the top row, and each house is an infinite sequence of Ys and Ns to represent which queens are and aren't members of the house.

	Q1	**Q2**	**Q3**	**Q4**	**Q5**	**Q6**	**Q7**	**Q8**	**Q9**	**Q10**	**...**
House 1	Y	Y	N	N	Y	N	N	Y	N	N	...
House 2	N	N	Y	Y	Y	Y	N	Y	Y	Y	...
House 3	N	Y	N	Y	N	Y	N	Y	N	Y	...
House 4	Y	N	Y	N	Y	N	Y	N	Y	N	...
House 5	Y	N	N	N	N	N	N	N	N	N	...
House 6	N	Y	Y	Y	Y	Y	Y	Y	Y	Y	...
⋮	⋮	⋮	⋮	⋮	⋮	⋮	⋮	⋮	⋮	⋮	⋮

House 1, which looks like YYNNYNNYNN . . . includes queen 1, queen 2, queen 5, queen 8, and so on. House 3 includes queens 2, 4, 6, 8, 10, and all the rest of the even-numbered queens, whereas house 4 contains only the odd-numbered queens. House 5 contains only queen 1, and house 6 contains every queen *except* for queen 1. There are infinitely many possible houses and, according to Cantor, this world of ballroom power sets yields a greater infinity than we've seen before! To see how, let's try to match each house with a single queen, as we've done before with wigs. After all, every house needs a mother (or father or parent) who leads it. Suppose we allow queen 1 to be the mother of house 1, queen 2 the mother of house 2, queen 5,000 the mother of house 5,000, and so on.

It would appear as though every queen is mother to a house, and every house has a queen to mother it, right? This is an illusion! At the level of infinity where we're now operating, there are so many possible houses that there aren't enough queens to be mothers to them all, and I can prove it.

In our scenario, a mother may or may not be a member of the house she's assigned to. For example, queen 1 is the mother of house 1 and also belongs to house 1. Queen 2 is the mother of house 2 but does not belong to the house itself. We can see, by looking at the diagonal line of the grid, that house 1 includes queen 1, house 2 does *not* include queen 2, house 3 does not include queen 3, and so on.

	Q1	**Q2**	**Q3**	**Q4**	**Q5**	**Q6**	**Q7**	**Q8**	**Q9**	**Q10**	...
House 1	**Y**	Y	N	N	Y	N	N	Y	N	N	...
House 2	N	**N**	Y	Y	Y	Y	N	Y	Y	Y	...
House 3	N	Y	**N**	Y	N	Y	N	Y	N	Y	...
House 4	Y	N	Y	**N**	Y	N	Y	N	Y	N	...
House 5	Y	N	N	N	**N**	N	N	N	N	N	...
House 6	N	Y	Y	Y	Y	**Y**	Y	Y	Y	Y	...
⋮	⋮	⋮	⋮	⋮	⋮	⋮	⋮	⋮	⋮	⋮	⋮

Since our list must eventually include all possible houses, what about the house formed by all mothers who are not members of their own house? Let's call this house *X*. House *X* would include queens like queen 2, queen 3, queen 4, and queen 5, who are not members of their own house, so house *X* would look something like this: NYYYYN. . . . Essentially, it is the exact opposite sequence of the diagonal line of the grid! Since house *X* is a house just like any other, it should have a mother somewhere on our infinitely large spreadsheet. Who is the mother of house *X*? Is she somewhere down the list? Perhaps the more important question is this:

Does the mother of house *X* belong to house *X*?

Remember that house *X* is formed by all mothers who *do not* belong to their own house. If the answer is yes, and the mother of house *X* does indeed belong to her own house, then the house rule dictates that she should *not* be a member. If on the other hand she is *not* a member of her own house, then the house rule dictates that she *must* be a member.

We're facing a contradiction. If house *X* has a mother on the list, then she either is or isn't a member of the house itself, but not both! The problem here is that house *X can't* have a mother. There simply aren't enough mothers to go around. As a matter of fact, there are infinitely many houses that we can construct which are motherless. The number of possible houses is simply a larger infinity.

Mathematicians call $\aleph_0$ *countable infinity*. We can think about countable infinity within the same context we use to enumerate things on a list because it's just like matching queens up with wig 1, wig 2, wig 3, wig 4, and so on. Even though counting an infinite number of queens or wigs one by one would take an eternity, in some sense this is possible because we can assign each a number, starting with 1, 2, 3, 4, 5, and so on, and be sure to include everyone with this labeling procedure. The cardinality of our infinite set of queens is the same as the cardinality of the natural numbers. They both have an $\aleph_0$ number of elements, which is countably infinite. However, the power set of the natural numbers, or the set of all houses of queens, is uncountably infinite. No matter how you rearrange them, there's no way to enumerate them all in a list; therefore we can't match them all up with mothers.

If the number of possible houses is larger than $\aleph_0$, then just how large is it? If the set of all queens has the same cardinality as the natural numbers, then the set of all houses also has the same cardinality as the *real* numbers. "Real" numbers include everything on the horizontal number line, from all of the whole numbers plus everything in between. In between 0 and 1 alone there are an infinite number of smaller and smaller fractions, like ½, ¼, ⅛, 1⁄16, 1⁄32, 1⁄64, and so on. There are even an infinite number of fractions between 0 and 0.00000000001. All these fractions, together with the whole numbers, are known as *rational numbers*. However, in between all the rational numbers are the irrational numbers we considered earlier, like π and the Champernowne constant. The set of irrational numbers is *uncountably* infinite! It is thanks to the uncountability of the irrationals that the collective family of real numbers is so large. We can label the cardinality of the real numbers with the letter $\mathfrak{c}$, to represent the continuum of the number line; $\mathfrak{c}$ is a transfinite number similar to $\aleph_0$, but even greater. We can use this transfinite number in equations just like we did with $\aleph_0$, but there's a new rule:

$$\aleph_0 + \mathfrak{c} = \mathfrak{c}$$

A smaller infinity plus a larger infinity is equal to the larger infinity. If there are an $\aleph_0$ number of rationals and a $\mathfrak{c}$ number of irrationals, then put together they comprise the real number continuum, which thus has a cardinality of $\mathfrak{c}$. The number of possible houses, the number of points on a continuum, and the number of points on a circle are all $\mathfrak{c}$. If gender is a spectrum, that means there would be at least a $\mathfrak{c}$ number of genders. Uncountably many! If there were an $\aleph_0$ number of stars in an infinitely expansive universe, $\mathfrak{c}$ would be the blackness between the stars.

The framework of sets and cardinality initially introduced by Cantor gives us a more meaningful way to talk about a concept we once thought was merely abstract.

You might argue, though, that everything you've read in this chapter has been abstract, from immortal monkeys to infinite queens. In real life, nothing really lasts forever or comes in an infinite quantity. But infinity and its different levels are researched and used prolifically throughout math. In probability theory, problems can be categorized as discrete or continuous, depending on whether the number of possible outcomes is countable (as in dice rolls or coin flips) or uncountable (as in time or distance, where there is a continuous range of values).

And as large as $\mathfrak{c}$ may be, there are greater infinities still. Remember that we started with an infinite set of queens, which had a cardinality of $\aleph_0$. The power set of the set of queens was our set of all houses, and that had a cardinality of $2^{\aleph_0}$, otherwise known as $\mathfrak{c}$. We can take the power set of this power set, which would be the set of all subsets of houses. That set would have a cardinality of $2^{\mathfrak{c}}$, which makes it infinitely larger than the previous set.

Just as there is no limit to the number of numbers we can encounter by simply adding together 1s, there's also no limit to the number of infinities. There are infinitely many uncountable infinities, with each one growing bigger and even more incomprehensible. Infinity reminds us that there are endless possibilities in what we can conceive and what we can become. It turns out that infinity is smaller than we thought it was, larger than we thought it could be, and queerer than anyone ever imagined.

CHAPTER 2

Celebrity Numbers

Glamour. Mystique. Controversy. And inspiration for a good story! All these are the hallmarks of a celebrity drag queen, but also a celebrity number. Like drag queens, celebrity numbers have gained notoriety by stretching perception and defying definition. Certain numbers just seem to captivate us, generating endless interest and buzz, even engendering arguments about whether they count as numbers at all.

Muse, what is a number? Which mathematical objects can be called numbers and which cannot? When do we decide a number has been definitively classified, and who gets to make that decision? To be able to define, label, and categorize things has long given people a feeling of comfort and control. As someone who has lived between different definitions my whole life, I've thought a lot about how defining something is related to accepting it. My experiences have allowed me to see a lot of parallels between our perception of numbers and our perception of gender. Both concepts have shifted in meaning over time, and while we explain them to children in simple terms, the reality is much more nuanced and radical.

A child may define gender based on outward appearances. They might say that men have short hair, are generally taller, and have deep voices, whereas women have longer hair, are usually shorter, and have higher pitched voices. They may also reference societally imposed differences: boys like the color blue and grow up to be strong providers; girls like pink and grow up to be caring mothers. But these static definitions are insufficient.

My gender presentation varies day to day, from boy to queen to something in between. Technically you could call me *gender fluid*, meaning that my gender varies instead of being rigidly fixed. That also means I frequently get stared at and receive comments such as, "What are you?" and "Are you a man or a woman?" from people who are trying to figure out what to make of me and how to treat me.

Many things we associate with womanhood, manhood, queerness, or straightness are contrived. If you've never felt trapped by them, you may argue that the definitions are fine the way they are. You may even think that these definitions come from nature or biology.

But appearances alone are not enough to understand a person's gender, and neither are body parts. For example, intersex people are those born with anatomies that don't fit the typical definitions of male or female. In a typical male, every cell has one X chromosome and one Y chromosome; in a typical female, every cell has two X chromosomes. But a person can be born with some cells that have two X chromosomes and other cells that have one X chromosome and one Y chromosome, or be born with genitals that appear ambiguous. Some people may be intersex without even realizing it because, in some cases, doctors perform surgery on newborn babies to "normalize" the appearance of their genitals without obtaining the parents' informed consent!

Pinning down a precise definition of gender and sex is no simple task, and the widely accepted gender binary falls short. Yet any perceived deviation from this binary can result in a person being dehumanized, prosecuted, harmed, or even killed.

In the Philippines, we speak a gender-neutral language called Tagalog (one of multiple languages in the archipelago). We use the pronoun *siya*, a gender-neutral version of "he" or "she," and the single word *asawa* (or *jowa*) to mean "spouse" (or "significant other") instead of the gendered terms "husband" and "wife" (or "boyfriend" and "girlfriend"). Even the term *bakla*, which is usually translated "gay," blurs the lines between gay, bisexual, and transgender, whereas in Western cultures these are typically seen as more distinct identities.

The concept of heteronormative marriage and gender roles in the Philippines came about through Spanish colonization. In precolonial Philippines, gender was a spectrum that went beyond the rigid binary of male and female, and those who were in-between were not only tolerated but celebrated and respected. We know this because the sixteenth-century Spanish missionaries who first colonized the Philippines wrote about the *bayog*, feminine men who dressed as, presented as, and were treated as women. These *bayog* were male shamans who wore their hair braided with gold and performed spiritual rituals that centered on nature, healing, and the afterlife.

Coming from a culture where effeminate men were cast aside as atrocities, the Spanish were astonished to find that the *bayog* were leaders in their communities. When the colonizers introduced Catholicism to the islands, they also introduced the idea that gender nonconforming people were abominations and sinners, which explains why modern-day Philippines is not a friendly place for queer people. Today, gay people in the Philippines are frequently seen as perverts and sexual predators, and many Filipino parents feel their moral, societal, and religious duty is to shun or even beat their children for daring to be queer.

But I know that there is room for people's minds to open and for definitions to change. Not only cultural history but the history of math has proven it so. Evolving definitions of numbers have not only been accepted but have proven to be the very reason mathematics as a field has progressed and helped to shape a better world. If we look for examples of positive change, we see forward-thinkers, rule-breakers, and creative solutions appearing throughout every culture and every discipline. So without further ado, I'd like to introduce three celebrity numbers that have defied definition, broken down our preconceived notions, and changed our minds. They stand as my mathematical examples that changing minds changes everything.

Zero the Hero

Before there was math, there was counting. One of the oldest mathematical artifacts found to date is known as the Ishango bone, discovered buried in volcanic ash in the Democratic Republic of the Congo. This 20,000-year-old bone tool has lines etched into it that some scholars think were added to improve the user's grip, but others believe were used to count things. Counting is certainly the first math toddlers do when they start to learn about numbers. It is the very simple and effective framework through which we teach children the foundations of math. But if we view numbers only as tools for counting, why would we bother counting zero sheep, or zero wigs? Why should a concept such as zero be thought of as a number, in the same league as mathematical royalty, like 1, 2, 3, or 100? One reason is embedded inside the number 100 itself: zero acts as a placeholder, without which we wouldn't be able to tell the difference between 13, 103, and 10,300. Alright, in fairness, we don't *need* zero to represent these numbers. The old-fashioned way of counting things was using body parts (like your fingers) or tally marks like those that may be carved into the Ishango bone.

Actually, you may already be familiar with a number system that doesn't use zero at all, which is Roman numerals. The Roman numerals for 1 to 12 are: I, II, III, IV, V, VI, VII, VIII, IX, X, XI, XII.

The Roman numerals for 4 and 9 look like IV and IX, which can be understood as "one (I) less than five (V)" and "one (I) less than ten (X)." The numeral for 20 looks like XX (two tens), and the numeral for 50 is a new symbol, L. Thus, the numeral for 40 is XL, meaning "ten (X) less than 50 (L)," and the numeral for 49 is XLIX. Larger numbers require new symbols: 100 is represented by C; 500 is represented by D; 1,000 is represented by M. But despite this symbolic diversity, there is no symbol for the number zero. This is because the cultures that originally used Roman numerals didn't believe zero was, in fact, a number. The closest they had was perhaps a capital N for *nulla*, or "none." The Greeks and Romans found the idea of "nothing" as a number to be as preposterous as a line with no length or a triangle with no area. Zero defied logic and their rigid perceptions of numbers. (Although their ideas about men's sexuality were notably less rigid.)

There is an important difference between numbers and numerals worth mentioning here. A *numeral* is not another word for a number; instead, a numeral is a symbol or notation we use to represent a number. The Roman V, the Hindu–Arabic 5, and the Chinese 五 are all different numerals that represent the same number. One downside of the Roman numeral system was that one would need to invent new notations for larger numbers. Elsewhere, the ancient Babylonians, Chinese, and Mayans developed what we call positional numeral systems that didn't have this flaw. Positional numeral systems can reuse and recycle the same set of symbols to represent new, larger numbers.

For example, the Babylonian symbol for 1, 𒁹, is the same as the symbol for both 60 and 3,600, and they relied on context to tell the difference. The number 2 was two 1s put together, 𒈫, which could also be 120 (two 60s put together). However, to represent the number 61, a space was required between the symbols, 𒁹 𒁹. That space evolved into a punctuation symbol that looked like two slanted lines, an early proxy for zero.

Elsewhere in the world, the Mayan civilization of Mesoamerica developed a similar placeholder symbol for zero. Their positional numeral system used a shell-like symbol to indicate zero, a dot for one, and a bar for five. As far as we can tell, the Mayans used numbers mostly for calendars, but our knowledge of their mathematics is extremely limited because most of their books and art were burned to ashes by Catholic Spanish colonizers who considered them to be the works of the devil. We may never know exactly how the Mayans developed their numeral system, or what they really thought about the nature of zero.

Zero had its use as a placeholder in numerous civilizations, but the one that made its way into the English language was India's zero, known then as *sunya*. In the seventh century CE, the Indian mathematician Braḥmagupta was the first to treat zero as its own number rather than simply a placeholder symbol. In particular, he wrote down the mathematical rules of working with zero: What happens if you multiply a number by zero? What if you add or subtract zero from a number? What if you divide zero by zero?

Indian mathematicians not only gave us zero but all ten digits from 0 through 9, which we now call the Hindu–Arabic numerals because they traveled through the Arab world before arriving in Western Europe. Arab mathematicians translated the Indian word *sunya* to the Arabic word *sifr*, which also meant "nothing"; then *sifr* made its way into Europe, where it was Latinized into *zefiro*, or simply *zero*. The Hindu–Arabic numerals gave us the ability to express an infinite number of numbers using only ten different symbols because each symbol takes on different meanings depending on where it's placed. Compared to the Roman numeral V which always means five, the Hindu–Arabic numeral 5 can mean 5, 50, 500, and so on, depending on its position.

Consider the number 595: the first 5 means 500, the 9 means 90, and the last 5 means 5. Read aloud it is "five hundred ninety-five." If we write 5,950, we recognize that the first 5 actually means 5,000, the 9 means 900, and the second 5 means 50. The full number is read as "five thousand nine hundred fifty." Each digit's position in the number represents a new power of 10, starting from 1, to 10, to 100, to 1,000 and so on. The number 5,950 is understood as having five "thousands," nine "hundreds," five "tens," and zero "ones." Mathematicians call this a base-10 numeral system.

The Babylonians used a base-60 system, meaning that each digit represented a new power of 60, from 1, to 60, to 3,600 and so on. Software programmers speak the language of binary numbers, which strictly uses 1s and 0s. Binary numbers operate on a base-2 numeral system, where each digit represents a power of 2: 1, 2, 4, 8, 16, and so on. In binary, the number two is encoded into the binary number "10," which signifies 1 lot of two and 0 lots of one. The binary number 110 signifies 1 lot of four, 1 lot of two, and 0 lots of one, which adds up to the (base-10) number 6. Computers use binary numbers, and the 1s and 0s can be translated into an on/off switch within an electric circuit. Our advancements in computers and technology are in part thanks to the heroics of zero.

Transformational and mind-blowing, pushing through limits and crushing boundaries, the elegant zero is simply iconic.

The switch from Roman numerals to Hindu–Arabic numerals made math significantly more efficient, especially when facing large numbers. That meant improvements in engineering, science, astronomy, architecture—you name it. However, zero wasn't initially met with open arms. Think about how strongly Americans resist switching to metric units instead of US customary units of measure, and now imagine trying to convince them to switch their entire numeral system to a completely foreign one that most had never heard of. It would require a lot of convincing! Even if successful, such a change could cause mass confusion and ample opportunities to commit fraud against people who struggle with math as it is. On top of practical reasons like those, there was an ideological push against the very concept of the Hindu–Arabic number system. For people accustomed to Roman numerals, accepting the abstract concept of zero as its own number smaller than 1 would also open the door to the abstract notion of negative numbers, and people didn't like or understand that one bit.

But they were wrong to resist! There was no stopping the great leap in efficiency that zero (and eventually negative numbers) would eventually bring to the Western world. Our understanding of numbers broadened into a number line that extended in two directions, with zero being the boundary between positive and negative numbers:

$$\leftarrow -5, -4, -3, -2, -1, 0, 1, 2, 3, 4, 5 \rightarrow$$

If accepting the "numberness" of zero was a hard pill to swallow, even more so was accepting the idea of negative numbers. When my teachers first introduced the concept of subtraction, with examples like 5 minus 3, or 6 minus 1, we weren't allowed to ask about 3 minus 5, or 1 minus 6. How could you take away 5 apples if you only started with 3? Yet once you learn a little about negative numbers in school, they start to feel natural. If 5 minus 3 is 2, then 3 minus 5 is –2. There's a natural symmetry when we reverse the order of subtraction. Today, we often see negative numbers representing basement levels in multistory buildings, frigid temperatures below zero degrees, and dollars of debt. Negative numbers expand our mathematical tool kit and our entire notion of the way numbers can represent reality.

But that's not all. Between each whole number are fractions, like a half, or a fifth, which we can represent in decimal notation as 0.5 or 0.2. Every number you can write as a fraction, or ratio between whole numbers, can be written in decimal, and vice versa. The number 1.999 can be written as 1,999/1,000, and –3.75 can be written as –15/4. We can make infinitely many fractions that live on the number line between the whole numbers. Some fractions lead to repeating decimals, like 10/3 = 3.33333 . . . , or 1/7 = 0.**142857**142857142857. . . . All fractions are *rational* numbers because they can be written as a *ratio*! This may be a very different definition of the word "rational" than people are used to. Whole numbers (positive, negative, and zero) are rational because they can be written as fractions: 1 = 1/1, –2 = –2/1 or –4/2, 10,000 = 10,000/1, and 0 = 0/1.

Are *all* numbers rational? Our exploration of infinity showed us that the answer is no. Our definition of the term "number" must once again expand to embrace the concept of an *irrational* number. And the most famous irrational number of all time is π.

The Digit Diva π

A lot of the clothes that I wear in drag are my own creations. I taught myself to sew because I got bored with the styles I was seeing in the ladies' section at H&M, and I couldn't afford anything from the ladies' section at Versace. I'm by no means a professional, but I make do with some stretch fabric and a safety pin. I have to admit I had no idea how much math was required in sewing until I began learning myself. Sewing requires precise measurements, matching together congruent patterns, and a knowledge of how to geometrically translate a fit from a three-dimensional body onto a two-dimensional sheet of fabric. Pi (represented by the symbol π) is a necessity every time I want to sew a dress with a circular skirt attached to it.

To cut out a circle skirt, you have to make a donut shape using fabric—a big circle of fabric with a smaller circle cut out from the center. To fit snugly on my body, the small circle has to be the same size as my waist, and given that my waist measures 85 cm all around, the challenge is to cut out a circle that is exactly 85 cm all the way around (this is called the circle's *circumference*). Drawing a circle freehand is not easy. You could trace around a circular object like a plate or a cup, but those circles have a fixed circumference that you can't alter to fit your measurement needs. The best way by far is to choose a point to be the center and then fold the fabric up around it, using that point as the corner, as if you were cutting out a paper snowflake (as my YouTube viewers have seen me do many times; see page 30). If you make a curved cut at the corner, you can unfold and see that a circle is magically cut out. The key is to know the size of the radius to use—the radius is the length from the edge to the center. The farther from the center you cut, the larger the circle. A tiny 1 cm radius will produce a waist-hole only big enough for a doll, and a big 100 cm radius would make a hole big enough to fit around a bed. Happily, I don't need to figure out exactly where my waist size falls between a Barbie and a Tempur-Pedic mattress because when it comes to circles of all sizes, π is the answer to our problems.

Many minds conjure up the mystifying number π when we think of next-level mathematics. This is a number so popular that it has its own annual holiday—putting it up there with Abraham Lincoln, Jesus, and the Queen of England!

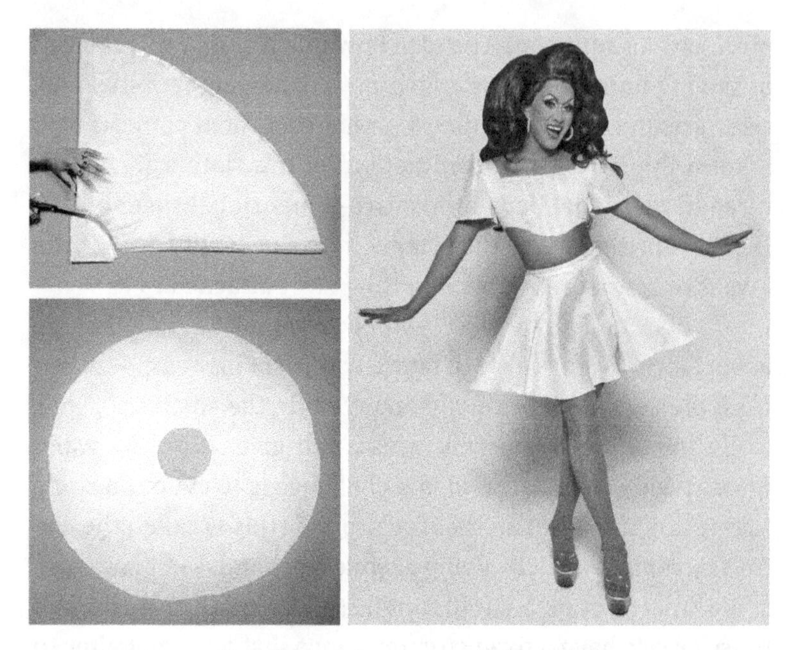

Kyne folds and cuts out a circle skirt.
Author photo by Fabian Di Corcia

Every March 14, we pay homage to this incomparable diva with memes, dad jokes, delectable desserts, and viral videos from yours truly! I'd go so far as to call π true mathematical royalty. The word "pi" comes from the Greek word *periferia*, which means the periphery, or circumference, of a circle. The number π is the ratio of a circle's circumference to its *diameter* (the distance across it, or twice the radius). I think our fascination with π may be thanks to the ubiquity of the circle in nature. There are circles in our eyes, in the sky, in ripples on the surface of a lake after being touched by a raindrop, and in a flower's bloom as its petals unfurl. No matter which circle you choose, if you measure its circumference and divide that number by its diameter, you will consistently get something close to 3.14, allowing for a small margin of error depending on how precise your measuring tape is.

Since π is approximately 3.14, we know that a circle with a circumference of 85 cm must have a diameter of about 27 cm (85/3.14). The radius, which is the distance from the edge to the center will thus be roughly 13.5 cm, since that's half the diameter (remember the diameter is the distance from edge to edge). If I cut 13.5 cm away from the center, I get the perfect hole to fit my waist.

But where does 3.14 come from? Archimedes of Syracuse, who lived around 200 BCE, devised a very clever way of calculating π that didn't rely on any clunky measuring tape. He simply used the rules of geometry on a hypothetical circle, which allowed him to use exact, theoretical values. Archimedes imagined a theoretical circle with a diameter of 2 and said the circumference must be $2 \times \pi$, whatever π may be. Then he drew a hexagon, a six-sided object, inside it and measured its perimeter, which was 6. Since the hexagon and the circle were close in size, Archimedes reasoned that $2 \times \pi$ (the perimeter of the circle) was close to 6 (the perimeter of the hexagon). Solving this equation would mean that π is approximately equal to 3.

Next, he added more corners to the hexagon by turning each straight edge into two new edges, resulting in a 12-sided polygon. By using the information about the hexagon along with more geometric rules (including the theorem of the controversial philosopher Pythagoras, $a^2 + b^2 = c^2$), you can work out that the perimeter of this 12-sided figure is approximately 6.2, which gives π a new approximation of 3.1. As Archimedes moves from a 6-sided figure to a 12-sided one, and then later a 24-sided, a 48-sided, and finally a 96-sided polygon, the polygons get closer and closer to the true shape of the circle, and the approximations of π get closer to the true value. Archimedes stopped by saying that π was somewhere between 3.1408 and 3.1429. I guess his hands got tired; after all, this was an age before calculators!

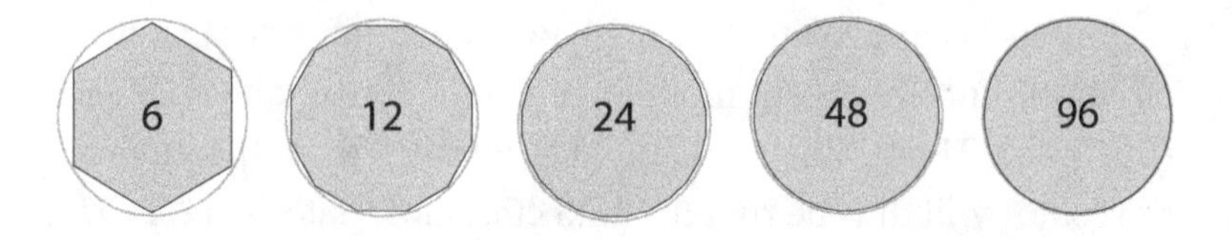

Archimedes turned what was once a matter of measurement into a matter of arithmetic, and a new age of mathematics was born. Now you might be saying, "But, Kyne, no matter how many sides a polygon has, it will never be a circle!" Which is exactly right, and that's why we will never get closure on this issue because we'll never find the exact value of π. It is an elusive, unknowable enchantress, with infinitely many digits. Mathematical paparazzi can chase her forever, but they will never capture her in her entirety.

If we are only concerned with using π to cut out dresses or build circular buildings, then we don't actually *need* to know infinitely many digits. We can get by perfectly fine with just the first few. In fact, even the scientists at NASA only use up to 15 digits and, wilder still, it only takes about 40 digits to calculate the circumference of the visible universe with a margin of error thinner than a hydrogen atom!

And yet, like desperate lovers chasing a dream of unearthly beauty, mathematicians through the ages have taken up the challenge, picking up right where Archimedes left off, investigating π on a journey of pure obsession. About 1,500 years after Archimedes, the Indian astronomer Madhava of Sangamagrama made a huge leap in calculating π, discovering that it could be calculated using this formula:

$$\pi = 4 - \frac{4}{3} + \frac{4}{5} - \frac{4}{7} + \frac{4}{9} - \ldots$$

The pattern starts with 4 (which is 4 divided by 1) and then alternates subtracting and adding 4 divided by the next odd number. Using only the first five terms, we can approximate π

to about 3.3397. If instead we extend the pattern to 10 terms, we get:

$$\pi = 4 - \frac{4}{3} + \frac{4}{5} - \frac{4}{7} + \frac{4}{9} - \frac{4}{11} + \frac{4}{13} - \frac{4}{15} + \frac{4}{17} - \frac{4}{19} = 3.0418$$

The more terms we use, the closer we get to the true value of π. Another 400 years after Madhava of Sangamagrama, the quest to understand and calculate π was turned upside down by the development of the first computer. The computing power of the digital age has progressed us from knowing a few dozen digits of π to knowing millions. Although the brilliant mathematician Johann Lambert proved in the 1700s that π was an irrational number with infinitely many digits that never terminate nor repeat, people persist in searching for more. I suppose it's for the same reason people try to break any sort of record. We do it for fun, for sport, or simply to test ourselves (or our computers).

Irrational numbers like π are special. With infinitely many nonrepeating decimal digits, they refuse to be written. To write down an irrational number, the Hindu–Arabic numerals 0 to 9 are not enough, nor can we use a fraction as we do with 1/3 or 1/7. We have to use mathematical equations to point at irrational numbers.

Another example of an irrational number is the square root of 2: 1.41421356. . . . Like π, the square root of 2 cannot be written down in plain digits. We settle for calling it the square root of 2 $(\sqrt{2})$. Unknowable like π, with that irrational mystique, $\sqrt{2}$ could definitely be nominated for celebrity status. And yet, it turns out that there are infinitely more irrational numbers than rational ones! If you threw a dart at a random spot on the infinitely long number line, you are almost certainly guaranteed to land at an irrational number and not a rational one. Irrational numbers like π are not the unicorns of the number world. If anything, it is the rational numbers, numbers like 1, 2, and 3, that are the rare gems. Once again, we see that the definition of a number—what we think we know for sure—stretches far beyond our preconceived notions. Numbers are more than just the discrete, whole quantities that we can count—numbers form a vast, continuous spectrum that extends infinitely in both directions: the real number line. But in the same way that I am asking you to envision gender as an infinite multitude of possibilities, I am going to ask you to let your imagination guide you into a place beyond the infinite number line. Because as a matter of fact, most numbers don't fit on the number line at all!

"But Kyne, where else can numbers live, if not on the number line?"

To begin to understand, we need to turn the spotlight on our final celebrity guest today. As the ultimate exercise in expanding our horizons, let's make some noise for a number almost as fluid as your hostess herself, the outrageous number *i*!

The Imaginary Number *i*

The lowercase *i* represents the "imaginary number," defined by the equation $i^2 = -1$. That notation means that *i* is a square root of −1. To give an example, 3 multiplied by 3 gives us 9, so 9 is called the "square" of 3, and 3 is a "square root" of 9. We use the language of squares here because for most of history Western mathematicians only thought in terms of geometric shapes. A square with a side length of 3 units has an area of 9 square units. This is the main reason why most ancient mathematicians resisted negative numbers—they couldn't fathom the idea of a square with a negative side length, much less a square with negative area. Essentially, *i* represents the side length of a square with an area of −1. The imaginary number *i* can't be a positive number, since two positive numbers always multiply to make another positive number. The same can be said about two negative numbers: −3 times −3 makes positive 9. On the other hand, zero is a number which is neither positive nor negative, but zero times itself equals zero. So what number could we multiply by itself to get −1? It seems that *i* must be a totally new type of number that we haven't encountered before. But just because we haven't encountered them, doesn't mean they don't exist. Just as zero was once unthinkable, and negative numbers nonsensical, and irrational numbers impossible, imaginary numbers were at one point equally ludicrous.

Let's travel back in time and try to put ourselves in the shoes of the frustrated mathematicians who stood at the edge of a breakthrough when they first encountered imaginary numbers. But first, a little bit of math.

Imaginary numbers first came in handy as a tool for solving equations. An equation is a mathematical sentence which has one equal sign and usually one variable, say x. Take for instance the following equation:

$$x^2 + x - 2 = 0$$

By the way, equations don't *have* to have variables in them. Here is a perfectly fine equation:

$$1 + 2 + 3 = 6$$

But these are less fun. If an equation has a variable in it, there's potential to *solve* the equation. Solving an equation that has a variable (x) means finding the particular value (or values) of x which make it true. One method to solve equations is by manipulating them with math. Around 825 BCE, the Arabian mathematician al-Khwarizmi worked out an algorithm for solving an equation with an x^2 like the one above (his name actually gives us the word "algorithm"), now known to high school students everywhere as the quadratic formula.

To determine the value of x in the general quadratic equation, $ax^2+bx+c=0$, use the quadratic formula:

$$x=\frac{-b\pm\sqrt{b^2-4ac}}{2a}$$

Applying his formula to the equation $x^2+x-2=0$ gives us the answers $x=1$ and $x=-2$, after substituting $a=1$ (there's an invisible 1 in front of x^2), $b=1$ (and another invisible 1 in front of x), and $c=-2$.

Another way to solve equations is with trial and error, by plugging in various possible values of x to see if they make the equation true. There are infinitely many possible answers, but you can sometimes get away with just trying a dozen, then plotting them on a graph and connecting the dots. With this equation, you'll get something that looks like this. The shape you see is called a *parabola*.

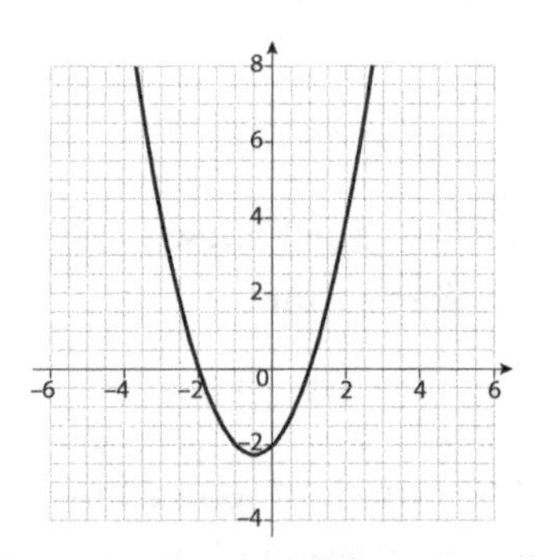

It turns out that solving the equation $x^2+x-2=0$ is equivalent to finding out where the parabola touches the x-axis, and it touches it at exactly where x equals 1 and x equals −2!

Now consider another example of a quadratic equation:

$$x^2+1=0$$

If we try to solve it algebraically, we might start by subtracting 1 from both sides:

$$x^2 = -1$$

The only value of x that could make this equation true would be one that, when squared, equals negative 1, or in other words, $x = \pm\sqrt{-1}$, or *i*. But this answer doesn't make much sense when we graph it:

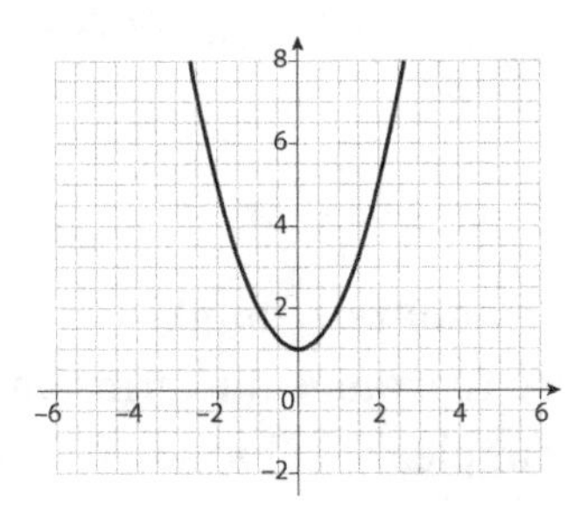

This parabola doesn't touch the *x*-axis at all! Some equations have no solutions. Or at least, they have no *real* solutions.

For centuries, mathematicians have enjoyed solving equations, just as detectives like solving puzzles. In Renaissance Italy, equation-solving was elevated to the point of igniting fierce math duels! These public duels would be fought between two prominent mathematicians, perhaps seeking new clients, new jobs, or higher salaries. Each mathematician would give the other a list of problems to solve, and whoever could solve more problems within a predetermined length of time, say about 40 days, would emerge the winner. These problems were hard but not impossible.

Challenging your opponent to solve a problem that you yourself could not solve was considered unfair, meaning that the best kind of problems were those that you—and only you—knew the solution to. This is in stark contrast to modern mathematics, where the solution to practically any mathematical problem is available at your fingertips, and if it isn't already known, mathematicians collaborate to find a solution and share any new progress with the community. But back in the world of math duels, mathematicians were reluctant to share anything at all. One such mathematician, Scipione Del Ferro, was the first person (to our knowledge) to discover how to solve equations like $x^3 + 2x = 5$. At the time, equations of this format (called depressed cubics) were thought to be unsolvable!

A *cubic equation* is any one that uses x^3, and possibly also x^2 and x. The most general form of cubic equation looks like this:

$$ax^3 + bx^2 + cx + d = 0$$

Here, a can be any nonzero number, and b, c, and d can be anything. We can also rearrange the terms on either side of the equal sign. Some examples of cubic equations follow:

$$x^3 + 2x^2 + 3x + 4 = 0$$
$$x^3 + 24x^2 = 3x$$
$$x^3 + 10x^2 = 5$$
$$x^3 = 0$$
$$x^3 + x = 0$$

All five equations are examples of cubic equations, but only the last two are called *depressed cubic equations* because they are missing the x^2 term.

The clever Del Ferro was sitting on a mathematical discovery that the greatest minds of his time hadn't worked out yet. But instead of sharing his discovery with the world, he kept it a secret to maintain his own job security. In the drag industry, queens sometimes refuse to reveal their makeup techniques or keep their costume designers' names a secret to stop any copycats from stealing their look. Some call these trade secrets; others call it gatekeeping! And Del Ferro was quite the gatekeeper. He wrote down his great mathematical discovery in his notebook, and only on his deathbed did he finally pass it on to a few trusted family members and his student, Antonio Maria Fior.

Del Ferro had been gone for a decade before his protégé, Fior, challenged mathematician Niccolo Tartaglia to a duel. Fior gave Tartaglia a list of 30 questions; all 30 were depressed cubic equations. He bet it all on his teacher's secret solution, hoping that his opponent wouldn't be able to uncover the same trick. But his hopes were dashed. Tartaglia solved all 30 equations. Although his victory spoils included a lavish banquet at the expense of the loser, Tartaglia magnanimously declined, and Fior sashayed away in disgrace.

Tartaglia had managed to independently discover the same solution Del Ferro found, making him the second person in the world to figure out how to solve depressed cubic equations. Like Del Ferro before him, Tartaglia refused to share his solution with anyone, leveraging it to reign victorious in all his future duels and insisting that he would publish the secret in his own book when he was ready. Tartaglia was like a drag queen with a pristine, unreleased version of Whitney Houston covering "My Heart Will Go On," an MP3 file powerful enough to sustain the rest of his career, and one that he'd go to great lengths to prevent anyone else from obtaining.

Tartaglia's secret solution might have been lost forever if not for a cheeky man named Gerolamo Cardano, who—whether through persuasion, threats of violence, bribery, or seduction we'll never know—eventually managed to convince him to give up the goods. Cardano swore to Tartaglia a solemn oath, pledging his faith as a true Christian never to reveal the formula, even going as far as promising to write it down in cipher so that no one could discover it after his death.

Despite his ardent assurances, some years later Cardano shared the secret with his student, Lodovico Ferrari, and the two of them expanded on Tartaglia's solution for depressed cubics, discovering how to use it to solve *all* types of cubic equations with an innovation that effectively reduced any cubic equation to a now solvable depressed cubic! From there, he'd let Tartaglia's algorithm do the rest and call it a day.

Cardano and Ferrari must have been thrilled to go where no other mathematical mind had gone before, but unlike Del Ferro and Tartaglia, they were uninterested in keeping this discovery a secret. Cardano was not a career mathematician who needed a weapon for engaging in math duels; he was after the clout. The only problem was that Cardano's sworn oath prevented them from publishing what they had learned. Even though they had made some original developments to Tartaglia's formula, their solution nonetheless relied on Tartaglia's work. Did Cardano have the right to share his discovery? Or did Tartaglia have the right to keep everyone in the dark? When I tell this story in my videos, my viewers usually side with Cardano. Many feel as though mathematical discoveries should belong to the world instead of benefitting only a chosen few. Especially discoveries as groundbreaking as this one.

Sadly unaware that they would find support for their decision on social media 500 years later, Cardano and Ferrari wrestled with their conundrum, but they ultimately found a loophole to their ethical dilemma when they got their hands on Del Ferro's original notebook. Del Ferro was the very mathematician whose penchant for secrecy started this whole saga in the first place! This meant Cardano could publish his solution and give posthumous credit to Del Ferro as the original author, technically bypassing his promise to his old friend Tartaglia. But believing he ought to give credit where credit was due, he also cited Tartaglia in his publications. Tartaglia reacted with horror. In his eyes, Cardano was a scoundrel who had violated a sacred oath. For better or for worse, we have those solutions because of him, so let's discuss and disseminate.

Cardano compiled all his works in a book titled *Ars Magna*, or *The Great Art*. In it, he covers how to solve certain cases of linear, quadratic, cubic, and even quartic equations (quartic equations are a step up from even cubics, allowing x to be raised to a power of 4). Here is Cardano's formula for the cubic equation, the one you've all been waiting for:

$$\text{if } x^3 + mx + n = 0$$

$$x = \sqrt[3]{-\frac{n}{2} + \sqrt{\frac{n^2}{4} + \frac{m^3}{27}}} + \sqrt[3]{-\frac{n}{2} - \sqrt{\frac{n^2}{4} + \frac{m^3}{27}}}$$

Remember when I said that Cardano figured out how to solve all types of cubic equations? Okay, I sort of lied. He figured out how to solve *most* cubic equations. Because all this happened before mathematicians embraced negative numbers. They were just beginning to adopt zero and the rest of the Hindu–Arabic numeral system, but negative numbers were still beyond their grasp. Accordingly, Cardano carefully avoided not only negative answers but also taking the square roots of negative numbers. And this is where Cardano ran into trouble.

He specified exactly which instance of cubic equations would involve stumbling into the dreaded square root of a negative number, and he called this *casus irreducibilis* (the irreducible case).

When working with quadratic equations, it was easy to avoid ever taking a square root of a negative number because that corresponded to a graph of a parabola which never touched the x-axis at all. Remember the equation $x^2 = -1$ discussed previously? This is one where, technically, we could try to solve by taking a square root of -1. But if we were to graph it, we'd see that the parabola doesn't even touch the x-axis, meaning, to avoid having to take the square roots of negative numbers, one could say the equation had no solutions. But a cubic equation *always* has a solution. Cubic equations don't take the shapes of parabolas, but rather they point up toward positive infinity in one direction and down toward negative infinity in the other direction. That means they have to cross zero at some point. Take a look at these three examples:

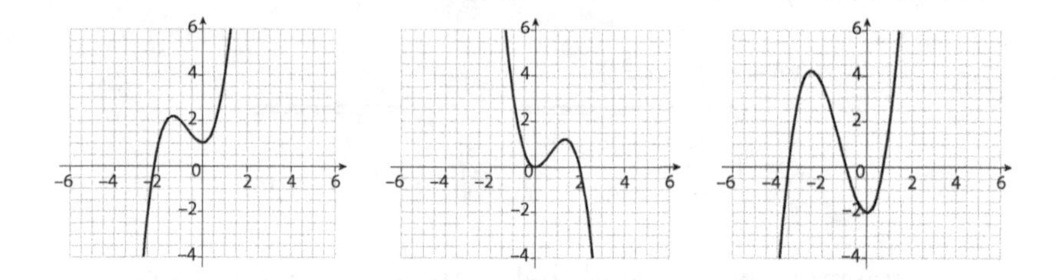

Cubic formulas *always* cross the *x*-axis, so Cardano couldn't just say that a cubic equation had no solution whenever he didn't like the math. When he restricted the parameters to avoid square roots of negative numbers, he was only able to solve cubic equations that crossed the *x*-axis once, as in the case of the first example above. Avoiding the *casus irreducibilis* proved to be a roadblock in fully understanding the nature of the cubic equation. It turns out that cubic equations with three solutions requires solving them with imaginary numbers. When Cardano occasionally includes an example that leans into imaginary numbers a little bit, he reacts with obvious dismay, calling these numbers "as subtle as they are useless." In Chapter 37 of *Ars Magna*, Cardano deals with the problem of finding two numbers that add up to 10 but multiply to make 40.

> If it should be said, Divide 10 into two parts the product of which is 30 or 40, it is clear that this case is impossible. Nevertheless, we will work thus: We divide 10 into two equal parts, making each 5. These we square, making 25. Subtract 40, if you will, from the 25 thus produced, as I showed you in the chapter on operations in the sixth book, leaving a remainder of −15, the square root of which added to or subtracted from 5 gives parts the product of which is 40. These will be $5+\sqrt{-15}$ and $5-\sqrt{-15}$.... Putting aside the mental

> tortures involved, multiply $5+\sqrt{-15}$ and $5-\sqrt{-15}$, making $25-(-15)$ which is +15. Hence this product is 40.[1]

Cardano is sailing into completely uncharted territory here! He found the two numbers which add to 10 and multiply together to make 40: these numbers were $5+\sqrt{-15}$ and $5-\sqrt{-15}$. If we ignore the frustration of taking the square root of −15 and treating this like a normal operation, we can add these quantities together like so:

$$(5+\sqrt{-15})+(5-\sqrt{-15})=5+5=10$$

Moreover, if we try to multiply these the traditional way, we have:

$$\begin{aligned} &(5+\sqrt{-15})(5-\sqrt{-15}) \\ &=25-5\sqrt{-15}+5\sqrt{-15}-\sqrt{-15}\sqrt{-15} \\ &=25-(-15) \\ &=40 \end{aligned}$$

Because $\sqrt{-15}$ ultimately disappears by the end, Cardano works with it (although begrudgingly). Even if we cannot accept the legitimacy of taking a square root of a negative number, it serves some real utility!

How are we meant to interpret a number like $\sqrt{-15}$, when it can't be drawn as a square, cube, or any other geometric shape? What are the rules for adding or multiplying these puzzling numbers? Can they even be called numbers at all? Some years later, another Italian mathematician, Rafael Bombelli, decided to investigate this mystery further.

Bombelli knew that these strange creatures, imaginary numbers, behaved differently from other numbers and therefore needed new rules. He showed off the new rules he devised in an amazing example, described by the following equation:

$$x^3=15x+4$$

Simple guess-and-check gives the solution $x=4$.

$$(4)^3 = 15(4) + 4$$
$$64 = 60 + 4$$
$$64 = 64$$

However, plugging the parameters of this equation into Cardano's formula would give us:

$$x = \sqrt[3]{\left(2+\sqrt{-121}\right)} + \sqrt[3]{\left(2-\sqrt{-121}\right)}$$

Cardano would have called this an irreducible case, since he didn't know what to do with $\sqrt{-121}$ here. It didn't easily cancel itself out like $\sqrt{-15}$ did in the previous example.

But Bombelli took it upon himself to reconcile how this monstrosity could possibly equal 4. He used some simple but clever algebra to demonstrate that $(2+\sqrt{-1})^3 = (2+\sqrt{-121})$ and $(2-\sqrt{-1})^3 = (2-\sqrt{-121})$, so that they simply add to get 4.

All this meant that if you were willing to accept some rules for working with these strange imaginary numbers, Cardano's formula actually worked in the *casus irreducibilis* where he thought his formula had reached its limit. The problem was that you had to abandon the geometric interpretation of squares and cubes and dive into the world of negatives and imaginary numbers. Many mathematicians were not willing to take this leap. Nevertheless, Bombelli at least forced mathematicians to take imaginary numbers seriously. Imaginary numbers, although still highly suspect, were now considered useful in solving equations. And once Bombelli opened the doors of perception, the next century was filled with many advancements in math that would help ease people into accepting these unfamiliar numbers, which eventually came to be known as "imaginary."

So if you've been able to open your own mind to the idea of imaginary numbers this far, let me guide you a step further because, despite the name we give them, imaginary numbers are just as real as negative numbers, fractions, and whole numbers! If you're having a tough time understanding, that's alright. It's not easy to change your frame of mind to accept new things. For most of history, mathematicians used an old-fashioned, rigid framework of numbers that interpreted numbers using geometry. Every number was viewed as a length of a line segment or a ratio between two line segments. Multiplying a number by itself was interpreted as finding the area of a square, formed by adjacent line segments of the same length. Multiplying a number by itself a total of three times could be interpreted as finding the volume of a cube. This framework had no way of explaining negative numbers, or numbers like zero or π, and there was even less room for imaginary numbers.

It's usually easier for us to insist that new things ought to change themselves in order to fit into our frame of reference. But as we're finding, the ability to be flexible about our frameworks—whether those we build around using nonbinary pronouns like they/them, or about nongeometric numbers like those deemed imaginary—is crucial.

A few years ago, a friend of mine came out as nonbinary, and started using they/them pronouns. For example, instead of saying that "he" or "she" is coming to dinner, I'd say that "they" were coming to dinner. Some family members found it hard to wrap their heads around this. Just as imaginary numbers seem to break established limits, but are very real nonetheless, transgender people may not fit into an old-fashioned, rigid framework of gender; but that doesn't make their existence any less real. You will continue to encounter new people throughout your life. Many of them will surprise you, or confuse you, or challenge your ideas. They may eat different foods, speak different languages, dress differently than you, pray differently than you, have different kinds of relationships than you're used to, or identify with labels you've never heard of and may personally find confusing. Mathematicians obstructed their own progress for hundreds of years by clinging to old limitations. We can learn from their mistakes. Because the first mathematicians who encountered imaginary numbers did not fully understand them, most chose to ignore them altogether. The ones who were willing to take a leap into the unknown changed the course of history. We are cheating ourselves if we try to force reality to fit our vision of the world; instead we can change our framework to encompass all sorts of people and ideas. As your hostess, I suggest we practice expanding our framework right within these pages because *i* is here, queer, and we better get used to it.

To wrap our heads around the reality of imaginary numbers, let's try visualizing all numbers on a number line, with 0 in the middle, and the negative numbers extending infinitely far to the left and positive numbers infinitely far to the right. We can make notches on the line at some of the whole numbers, like –4, –3, –2, –1, 0, 1, 2, 3, and 4. We can imagine an arrow pointing to each positive number on the right side and an arrow pointing to each negative number on the left. Adding two numbers is like stacking those arrows together, and multiplying two numbers is akin to scaling an arrow. For example, multiplying 2 by 4 is like quadrupling the length of the arrow that points to 2 as it scales up to 8.

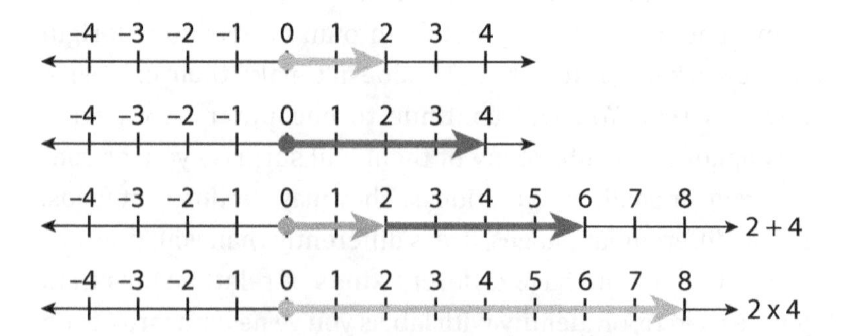

Multiplying by a negative number is like making a 180-degree turn—it switches positives into negatives and negatives into positives. What happens when we multiply by i? Multiplying $i \times 1$ is simply i (1 times anything gives you the same thing you started with). Now let's multiply by i again. What is $i \times i$? Remember that i is a square root of -1, so i^2 is by definition -1. Multiply this answer by i again: $-1 \times i$ is $-i$, so now we know that multiplying i by itself three times gives us $-i$. Multiplying this by i a fourth time gives us $-i \times i = -i^2 = -(-1) = 1$, which is right back where we started. We can summarize this with exponents:

$$i^1 = i \qquad i^2 = -1 \qquad i^3 = -i \qquad i^4 = 1$$

If we continue to calculate i^5, i^6, i^7, and i^8, we just keep repeating these same four values: i, -1, $-i$, and 1. Two successive multiplications of i turns 1 into -1, which is like a 180-degree turn, and four multiplications of i turns 1 back into itself, which is like a 360-degree turn. So we can interpret a single multiplication of i as a 90-degree turn counterclockwise onto a new *complex plane*, with the ordinary number line taking the horizontal axis and the imaginary numbers along the vertical axis.

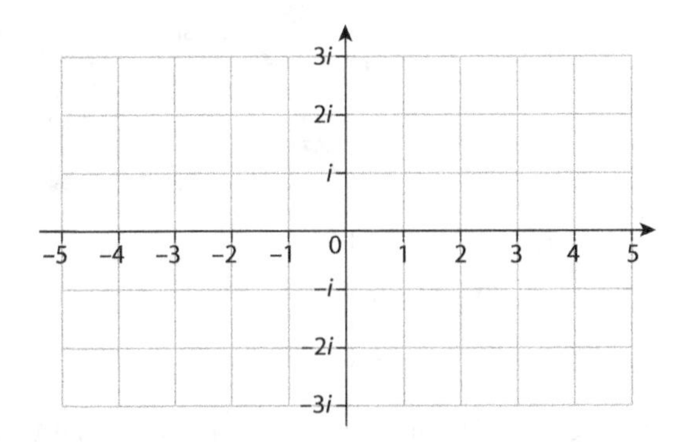

The complex plane.

Instead of living on the number line, imaginary numbers occupy a whole new axis, perpendicular to the traditional number line, meeting at zero! Since we now need a new name for the ordinary number line to distinguish it from our new imaginary line, we call the old (horizontal) number line the *real numbers*. But again, don't let these names trick you. Both real and imaginary numbers are equally legitimate. These are, in some ways, arbitrary definitions, not indicating true existence or lack thereof; a queen by any other name would look as sweet!

Once we see a whole new direction to our boring, straight number line, it prompts the question: What lives in the empty quadrants that our now crossed lines create? If we have real numbers living on the horizontal axis and imaginary numbers on the vertical one, what kind of numbers live in the rest of this space? These would be called *complex* numbers, and they are part real and part imaginary. You get a complex number by adding a real number with an imaginary one, such as $1+i$, or $2+3i$.

To add together complex numbers, you simply add together the real parts (usually written as the first term of a complex

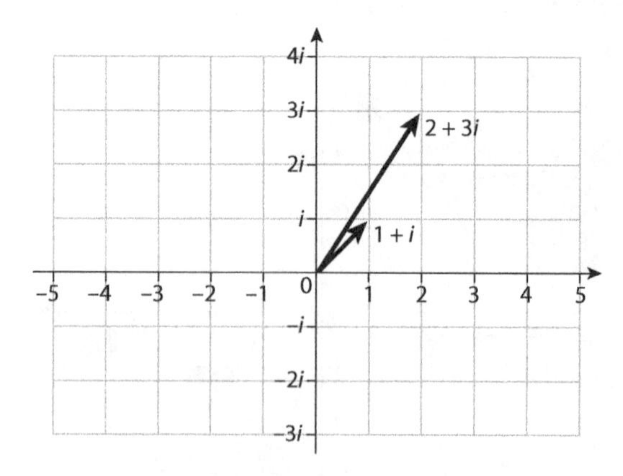

Two numbers that live in the complex plane, $2+3i$ and $1+i$.

number) and then add together the imaginary parts (usually written as the second term, and with an i attached to it). For example, $1+i$ plus $2+3i$ makes $3+4i$ (1 plus 2 makes 3, and i plus $3i$ makes $4i$).

On the complex plane, adding together complex numbers works the same as with regular real numbers: we attach the end of one arrow to the start of another, and what we get is two sides of a triangle, and the answer is the third side!

The analogy of rotations and arrows works perfectly, actually, because adding together complex numbers can be seen as stacking together two arrows on the grid, and multiplying them can be seen as adding together their rotations! Cardano couldn't have imagined how this application of imaginary numbers would come in extremely handy for electrical engineers, who use the math of rotations and multiplications to easily handle circuits and signals.

Every time Cardano had to deal with square roots of negative numbers, he thought that his formula failed, but using

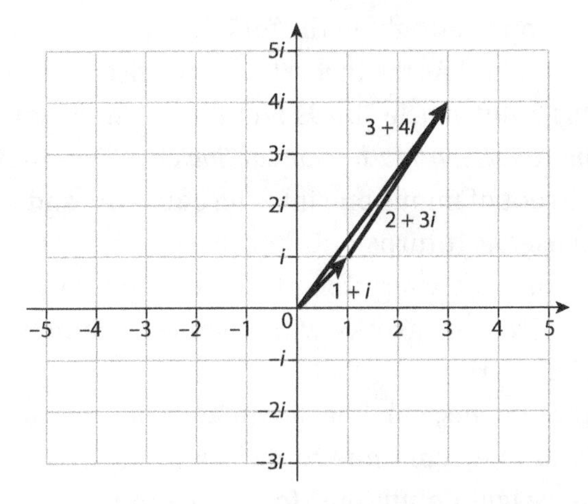

When we add together two complex numbers, it's analogous to stacking their arrows one after another. Their sum is the third side of the triangle.

imaginary numbers actually simplifies and expands the entire mathematical tool kit. Now we know that every single cubic equation has exactly 3 solutions, and some of them might just be complex numbers. Similarly, we can now say with confidence that a quadratic equation always has 2 solutions, a cubic equation has 3 solutions, a quartic equation (one that features *x* to the power of 4) has 4 solutions, and so on. This discovery is important enough that it was given a fancy name: the fundamental theorem of algebra! Not only do imaginary numbers help in solving particular equations favored by math duelists, but this new framework opens up a new world of possibilities. Imaginary numbers even make an appearance in the famous equation for quantum mechanics penned by feline aficionado Schrödinger:

$$\hat{H}|\psi_n(t)\rangle = i\hbar\frac{\partial}{\partial t}|\psi_n(t)\rangle$$

We're not dealing with hypothetical equations from math duels anymore; this is an equation that describes our physical universe, and yet right in the middle is *i*, the square root of a negative number! As the theoretical physicist Freeman Dyson put it in his 2009 article "Birds and Frogs," in the journal *Notices of the American Mathematical Society*: "Schrödinger put the square root of minus one into the equation, and suddenly it made sense. . . . It turns out that the Schrödinger equation describes correctly everything we know about the behavior of atoms. . . . [N]ature works with complex numbers and not with real numbers."[2]

Just like humans, numbers are unique, mysterious, and full of surprises. They continue to defy expectations and break through our rigid definitions, forcing us to rethink what we thought we knew. Numbers can teach us a lot about ourselves. Just as the discovery of imaginary numbers created a need to label the non-imaginary numbers, the coining of the word "transgender" created a need for a word that meant "nontransgender." That name, in English, became "cisgender." The prefix *cis* comes from Latin, meaning "this side of," and cisgender refers to people who identify with the same gender assigned to them at birth. The prefix *trans* means "the other side of," reflecting that trans people identify with a different gender than the one assigned to them at birth. The practical need for a new definition like this doesn't stop people from resisting it, though. Across the world you'll find people who are irritated by being labeled "cis." But here's the thing: cis women are women, just as real numbers are numbers; trans men are men, just as imaginary numbers are numbers. We saw that it took some time for humanity to accept the legitimacy of zero, and yet it's a concept so fundamental to us now that most people never dream of questioning it. It may take more time for us to accept imaginary numbers, and many people are still working on accepting the identities of trans people.

We may never stop discovering new types of numbers and new types of humans. As we encounter them, we may find that these beings are not new at all, and that they've been there all along—we just didn't have the accurate vocabulary to name them. Our frameworks will always need tweaking and adjusting, but the knowledge we gain along the way is worth the extra effort it takes to open our minds. And sometimes it takes a celebrity, whether numerical or in full drag, to show us how.

CHAPTER 3

How to Cut a Cake and Eat It Too

My stilettos first graced the stage thanks to a little show called *RuPaul's Drag Race*. If you haven't heard of it by now . . . where have you been? *Drag Race* is a reality show featuring roughly a dozen drag queens, who compete in comedy, acting, and fashion challenges. One queen gets eliminated each week until a winner is crowned America's Next Drag Superstar. At the time I discovered the show, I didn't have the best impression of drag queens. To be honest, I thought that they were just gay men who worked in bars, impersonating women, and I felt I had nothing in common with them. Even though I was on my own journey of self-discovery and enjoyed experimenting with makeup, I wasn't interested in flaunting long hair, breasts, or high heels. That all felt unnatural to me. I didn't want to look like a girl, and I didn't want people to refer to me as "she/her." I just wanted to be artistic and androgynous.

But then in high school, I started seeing loads of memes about the show, and they piqued my interest. I began streaming it from season 1, and the performances and personalities gradually opened my mind. While some queens on the show embraced the female impersonator label, others identified as androgynous, like me. Other queens on the show have identified as gender fluid, nonbinary, trans men, or trans women.

In the public imagination, drag queens are all towering, rambunctious creatures of the night. The wonderful thing about *Drag Race* is that it highlights the real human beings underneath the drag. I had never even been to a gay bar or met a drag queen before watching the show, but every episode humanized them for me. It introduced me to queens who were shy, sensitive souls from small towns, as well as bubbly, witty queens with a million talents. And perhaps most importantly, I started to see drag as a creative type of performance art. Soon, I was buying wigs to dress up as different characters, like Ursula, Cruella de Vil, or Marilyn Monroe.

When I attended my first drag show in person a few years later, I became convinced that I had to try out performing, myself. Watching drag on TV was nothing like experiencing it in person. The music, the dancing, the crowd interaction—I was hooked! Thus my drag career began, but not under a stage name because performing felt like the natural progression of my artistic hobby, not the birth of a new character. I liked that my real name sounded unique and androgynous already, so the first time I stepped on stage, I had them introduce me as me: Kyne.

When I started performing in drag shows across the province, I met all sorts of performers. Gay men impersonating women were just the tip of the iceberg, darling! I performed alongside female drag queens, drag kings, and drag things of all gender expressions—artists and misfits who didn't fit into any box. I never imagined that one day my drag career would lead me back to the show that set me on this path from the very beginning.

Years ago, *Drag Race* was strictly an American show, and us Canadian queens never thought we could be a part of it. But when the spin-off series, *Canada's Drag Race*, was announced, there was no question that I had to audition. I needed to see for myself where I'd land on the race track. A few months after sending in my audition tape, I received a call saying I had been cast on the inaugural season!

I jumped around the house in ecstasy.

Then the gears started to turn. . . . Who else is going to be there? What will the challenges be like? What kinds of outfits do I need to bring?

I spent the next few weeks preparing as if I were studying for a math exam. I rehearsed jokes with my husband that I could use for the improv challenges, and I practiced dance moves for the lip syncs. I spent somewhere around $6,000 on new costumes—from fabrics to sew my own dresses, to ready-made garments from cheap fast-fashion stores. I didn't have the guts to ask for help from professional designers or fellow queens; I was strictly instructed to keep my casting a secret, and I was such a stickler for rules that I refused to tell a soul, even to better my chances. When the scholarship coordinator at my university reached out to me to ask why I was taking a semester off of school, I told her I was pursuing a job opportunity. It wasn't a lie!

As a superfan of the show, I was convinced that there was more than one game I had to play. See, there's the main drag competition that people watch on TV, where the queens are judged based on how well they perform in the challenges and sashay down the runway. But there's another game going on behind the scenes: the show's producers are crafting an entertaining reality show. These hidden players watch the audition tapes and talk to the contestants during the confessionals. They're looking for queens with interesting (and sometimes tragic) backstories, who give memorable sound bites and demonstrate vulnerability.

Finally, there's a third game afoot—a competition to win over the viewers at home. Fans might not realize it, but their support is incredibly important because queens who emerge as fan favorites have the best chances to build successful careers as touring performers. By the time filming began, I was consumed with strategizing about how every one of my decisions might be interpreted by the judges, the producers, and the viewers alike. I wanted to do whatever it took to win, and to me, that meant having the most mathematically sound strategy.

The Games Divas Play

I was thinking particularly about my university course on game theory, which is an entire branch of math concerned with the strategy of decision-making. Game theory came into its own in 1928 when John von Neumann published a paper, "On the Theory of Games and Strategy," in which he posed the question: In a game of strategy with multiple players, how should one of the participants play in order to achieve the most advantageous result?

Game theorists like to talk about the best strategies for winning games like tic-tac-toe or chess. Chess is a lot more complicated than tic-tac-toe, but the basic principles are the same: try to anticipate what your opponent will do and outsmart them. The twist is that your opponent is also trying to anticipate what you will do so that they can outsmart you! We all know that in tic-tac-toe, the goal is to mark three boxes in a row, but as soon as you cross off two boxes, any savvy opponent will block you from marking the third. Many of us discovered as children that a winning strategy would be to try to stake your claim in two different rows at the same time; since your opponent can only choose one to block, you have ensured an opening to claim your victory.

Game theory applies to a lot more than just fun diversions. The "games" that game theorists study can refer to any real life scenario where there are multiple participants strategizing to achieve the best possible outcome for themselves. You're playing a strategic game when negotiating a salary, placing an offer on a house, or launching a business. Even driving home from work is a game. The other players are all the other vehicles and pedestrians on the road, and each one of you is seeking the best route to get to your destination as fast as possible.

When we think about games, most of us think about winners and losers. In a zero-sum game, in order for one player to gain a positive outcome, the opponent must suffer a negative outcome, unless you call a draw. But this isn't a very optimistic way to look at your life decisions! Many situations we find ourselves in *aren't* zero-sum games, and strategies exist that can make everyone happy.

One example of game theory that I love uses the analogy of cake-cutting. Suppose we have one cake to share between two people—let's call them Kelly and Michelle. Both of them want as much cake as possible. So how can Kelly and Michelle cut their slices fairly?

One option is to use a scale or measuring tape to cut two pieces with the same dimensions or weight.

But there's another method that involves nothing more than a cake knife: let Kelly cut the cake into two pieces and let Michelle choose which piece she wants, or vice versa. Since this is a strategic game, each is anticipating what the other will do. Kelly knows that Michelle wants the most bootylicious slice, so Kelly is incentivized to cut the cake as fairly as possible. She knows that if she cuts it in a way that makes one slice obviously bigger than the other, Michelle will just take the bigger piece. Kelly should therefore cut the cake so that she'd be equally happy with either slice. Once Michelle chooses the piece she thinks is best, Kelly gets whichever is left. Neither one can complain because Michelle got to choose which slice she thought was better, and Kelly set herself up to be equally happy either way.

What if we extend this thought experiment by adding a third diva to the group? How does our strategy need to change? If we let Kelly start by cutting the cake into three pieces that she deems equally fair, then between Michelle and our newcomer (say her name is Beyoncé), who gets to choose first? Whoever chooses second might be jealous, so let's use a different approach.

The cake can be any shape, but for the sake of easy visualization, let's imagine it's round like a clock. Once again, we'll start with Kelly making the cuts, but instead of splitting the cake into three slices, Kelly will make the first cut from the center to the edge at the 12 o'clock position. Next, she'll slowly move the knife above the cake, hovering over the cake's 1 o'clock, which would be one twelfth of the cake, then 2 o'clock, which would be a sixth, then 3 o'clock, which is a quarter of the cake, all while waiting for someone to shout, "Stop!" Kelly, Beyoncé, and Michelle are all free to speak up once the knife is hovering over their ideally sized slice. Kelly will cut the cake exactly where she's been stopped, and whoever spoke up first will get that slice. If more than one player shouts at the same time, they can break the tie however they want, maybe by flipping a coin. After the most vocal diva walks away with her slice, the remaining two girls will split the rest of the cake using the strategy we developed for the two-person cake-cutting scenario. If all goes well, each hungry diva will get about a third of the cake.

What if Beyoncé is feeling greedy and wants more than a third? She could try to hold out for a larger slice, say, a 40% share, but if she waits too long to say "Stop!" someone else might speak up first, like around the 36% mark, which would leave the queen with a lesser share. This would make for an unhappy Beyoncé, which frankly none of us would allow. But we needn't worry. Beyoncé knows that the leftover 64% will be cut by the "one-person cuts, one chooses" method, which does a pretty good job of ensuring a 50–50 split. Queen Bey is also fully aware that half of 64% is less than 36%, so she knows good and well that she better speak up as soon as the portion approaches 33.3%.

This strategy also stops overeager participants from speaking up too soon if they want a fair share. If any player hollers when the knife is only over 20% of the cake, then that leaves 80% to split between the other two. In our cake-cutting scenario, everyone is instead incentivized to wait until the knife is over the 33.3% mark (or at least at the place where the players *think* it's 33.3%) before calling a halt.

What if we added a fourth member, LeToya, to the party? All we have to do is use this moving-knife method again, having one person shout when they think the knife is hovering over a quarter, and leaving three people to split the rest. From there, we know how to reduce the three-person problem down to two people, and then we can bring it home using the "I cut, you choose" method again. There is a lesson here beyond just how to divide resources fairly—this method also demonstrates that if we know how to take a big problem and reduce it to a smaller problem, we can solve any conundrum, no matter how complex! We can generalize this strategy to any number of people fighting over any kind of cake. Imagine the ensuing chaos if the cake is asymmetrical, and has sprinkles and frosting more concentrated on one side than the other! And what if the cake cutter, Kelly, likes sprinkles a lot more than Michelle does? In this case, it may not be possible to avoid envy without changing our tactics somehow, but luckily game theorists have come up with clever new strategies for just about every "cakeventuality."

My favorite is ice cream cake, which could cause some cutting complications if it's left out too long, but let's leave that sticky issue aside because, of course, these algorithms are applicable far beyond cake.

Fair Valuation

My mom is the youngest of nine siblings, who all grew up in the Philippines on a plot of land a few acres in size. Recently, they all drew up plans to fairly divide the property into nine equal parts, but the land was much more oddly shaped than a birthday cake. Some of my uncles had houses on the land, so their share had to include their homes, and then there was a debate over who would get the piece of property that included my grandma's house. Many things, like houses, family heirlooms, artwork, or copyrights, are not easily divisible. None of these can be cut in half with a knife. Obviously there's the option of selling things and converting them to cash, which *can* be evenly split, but what about sentimental items, like a cross-stitched portrait or an old rocking chair, which may not have much value in a pawnshop but are priceless to the family members?

I think most families are willing to put up with a little bit of inequity to make the process run smoother. But how much unfairness would you be willing to accept if, instead of dividing up a few acres, you had to divide an entire country? After World War II, Germany was to be divided into four geographic regions, to be controlled by the United States, the United Kingdom, France, and the Soviet Union. The division ended up being loosely based on geography. France, which shares Germany's southwest border, got the southwest corner. The Soviet Union, the only country to Germany's east, was granted the eastern region. The United Kingdom, which was closest to the northwest corner of Germany, was given control over the northwest, and the United States was given the southern part of the country. It was decided that Berlin, the capital city, should be split into four regions. These lines in the sand decided the fates of millions of Germans for the next four decades; citizens either grew up on the Western, capitalist side allied with the United States, the United Kingdom, and France, or on the Eastern, communist side, which remained allied with the Soviet Union. Today, long after the physical wall has come down, differences in politics, religion, and wealth remain.

Let me give you a final example of ways to divide things fairly that's probably a little closer to home. Suppose that you and a friend have decided to become roommates, and you're moving into a two-bedroom apartment with a total rent of $2,000 per month, but one room is bigger and has an en suite bathroom. How do you decide who gets each bedroom? You could flip a coin to decide, but whoever gets stuck in the smaller room may end up feeling like they got cheated. Unlike a cake, which can be cut anywhere along its circumference, we can't cut into the walls of the apartment to resize the rooms. But something we *can* adjust in this scenario is how much we decide each room is worth. Instead of splitting the rent down the middle, you could agree that whoever gets the bigger room will pay a larger share of the rent. One way to decide the exact rent for each roommate is to base the rent on the square footage of each bedroom. If the bigger room is 50% larger, maybe that roommate should pay 50% more.

But what if, instead of size, one room has a better Wi-Fi connection, and another room has better natural lighting? How should we put those factors into numbers?

If we apply the lesson of "divide and choose" here, we can simply ask one roommate to state a division which they personally believe is fair, and let the other roommate choose. Suppose the first roommate believes that the big room with the en suite is worth $1,300 and the other room is worth $700, and they would be equally happy with either room at those prices. The second roommate is then allowed to pick which room they want at those prices, and the first roommate moves into whichever room is left over. Both roommates should then be satisfied with whatever room they end up in.

Now, what if we introduced a third roommate and a third bedroom? We can use a variation of the slow-moving knife strategy, but without tearing down any walls. We can essentially auction off one of the rooms: take any of the three rooms at random, but instead of starting the bidding at some low number, start the bidding at the maximum number. That's because when it comes to cake, we assumed that all the players wanted to *maximize* their slice, whereas roommates typically want to *minimize* their rent. So we can auction off one of the rooms at the maximum price of $2,000—the total rent. From there, decrease the rent, dollar by dollar, and each of the three roommates can shout once the price reaches an amount that they think is a fair valuation for that room. The first person to "say when" will get the room at that price, and the remaining two rooms can thus be split up by the remaining two roommates using the "I pick, you choose" method.

Fair valuation is one of the main areas of study among game theorists, so it makes sense that they have intensely studied the math behind auction strategies. Auctions don't just exist in fancy art houses or charity galas. Drag artists bid for wigs from stylists or retired queens, farmers bid for livestock, production companies bid for movie rights, oil companies bid for drilling rights, and telecommunication companies bid for bands of airwaves. Google's $200 billion yearly revenue comes mostly from ads that are auctioned: every time you type a term into their search engine—whether you are looking at wigs, cakes, or cows—advertisers are bidding to be put at the top of your search results.

Auctions are games too. The players are the bidders who all want the same item but to different degrees. Some may want it more than others, but everyone wants to minimize the price they pay. There are a couple of different ways to auction off an item. The method of auctioning off a bedroom by starting at a high price and slowly going down is called a descending-price auction, also known as a Dutch auction. You might use this method if you're trying to sell some old patio furniture on Facebook Marketplace: you start by listing a high price to see if anyone bites and, if no one's interested, you lower your asking price until someone is willing to pay. This is often referred to as "best offer" in the United States.

If you did the opposite and started at a low price and slowly raised it, you'd have to deal with messaging a bunch of interested buyers, who would probably get annoyed with you for raising the prices on them over and over again. The Dutch auction is much faster and thus is used to rapidly auction off things, like the freshly cut flowers from Holland that give the Dutch auction its name. In fact, approximately $400 million worth of flowers are sold this way every year in the Netherlands.

The ascending-price auction, also known as an English auction, is the kind you see in movies, where an auctioneer stands at a podium and shouts, "Going once, going twice, and sold!" as he slams his gavel. Aside from being longer, and more ripe for drama, than Dutch auctions, English auctions have another major difference—the opportunity to exploit the information gained from other bidders. Suppose a 1980s lunch box featuring comic book characters is up for grabs at an English auction. You don't recognize the characters, and the lunch box looks worthless to you, but someone bids $20. Shortly after, someone else bids $40. Do they simply like the way it looks? Could it be an antique? Next, someone bids $80. Does everyone know something you don't? Is this lunch box some sort of precious collectors' item?

After seeing everyone else's interest, you're tempted to join in, so you raise your hand and bid $100 on something you initially thought was worthless. Now, the other bidders don't know that you are completely clueless, so they may interpret your $100 bid as further confirmation that this lunch box has some real value, encouraging them to raise their bids in response to yours. You still don't know if the other buyers are lunch box connoisseurs, or if they are just as ignorant as you are, and yet here you are in the middle of a bidding war with them.

Dutch auctions suffer from the opposite problem. In a Dutch auction, the gavel drops after the first bid is placed, so there's no chance for you to hear what anyone else was going to bid. The question isn't how high the other bidders are willing to go, but instead how long they're willing to wait before making the first (and last) move.

Let's say you're bidding for a vintage Dior dress, and your budget is $5,000, but you'd prefer to set a little aside for matching shoes. If the seller's opening price of $8,000 quickly decreases, when should you jump in? It makes no sense to jump in while the prices are still above your budget, but what happens when the seller reaches $5,000? Should you jump in right away or wait it out and try to save some money? For all you know, there's only one other person with their eye on this dress, and their budget is $3,000. In theory, you could wait until the price has gone down to $3,001 before swooping in. But this is a risky game. You don't really know what anyone else is willing to pay for the dress, and cutting it too close might cause you to forever mourn the Dior that got away. According to game theorists, the optimal strategy is to bid just below your budget. If you'd regret losing the chance to buy the dress, you should speak up as soon as the price gets below $5,000.

Game theorists offer similar advice for the ascending-price English auction. Don't engage in bidding wars; bid at your budget and no more than that. The optimal strategy is the simplest one.

However, there is another (although illegal) strategy: collusion. If all bidders work together, they can find out which one of them is willing to pay the most for the item; therefore that person can be the winner without paying more than they need to. In the real estate market, homes are usually sold using a third type of auction: a blind auction, where there's a suggested asking price for the home, and then buyers privately send in written bids. In hot markets, these bids can be hundreds of thousands of dollars higher than the original asking price. Suppose you are bidding for a house and your maximum budget is $300,000. Somehow you manage to find the only other prospective buyer, and you want to collude with them so that neither of you have to bid over the asking price. The other buyer tells you that they are willing to bid $400,000 for the home. You now have a couple of options. You could tell them the truth and say that you're only bidding $300,000, allowing them to adjust their bid down to $310,000, giving them the house for a cheaper price. A more sinister option is to lie—you could *say* you're bidding only $250,000, tricking them into lowering their bid below $300,000, but then you *actually* bid $300,000. If they lowered their bid to $260,000 based on your disinformation, you would get the house. All's fair in love and residential real estate!

Searching for Equilibrium

A great example of collusion and double-crossing run amok plays out on an old British game show I particularly like, called *Golden Balls.* In the game, two contestants work together to win a cash prize of as much as £75,000. At the end of the episode, they are given the choice to either split the winnings in half or steal the entire amount for themselves. If one player chooses to steal and the other chooses to split, the one who steals gets to keep the entire prize, leaving the other with nothing. But if both players choose to steal, they both leave with nothing. We can write down all the players' options using a table like this, called a *payoff matrix*:

Payoff matrix for *Golden Balls*

	Opponent splits	Opponent steals
You split	You receive £37,500 They receive £37,500	You receive 0 They receive £75,000
You steal	You receive £75,000 They receive 0	You receive 0 They receive 0

The worst outcome on the board is if both of you walk away empty-handed, and yet it happens in about 25% of the episodes. The most frequent outcome is that one contestant steals and the other splits. Before they make their decision, the two contestants speak to each other on camera, often exchanging affirmations like, "I promise to split," "I'll split if you split," or "It's not in my nature to steal." If we analyze this as a game of strategy, where both players are only concerned with maximizing their winnings, then the best strategy is to steal. If your opponent splits, then you stand to double your winnings by stealing; whereas if your opponent steals, you will receive nothing no matter what you choose, unless you also choose to steal to be vindictive. In either case, choosing to split never benefits you. This game is a classic example of a "prisoner's dilemma," named after a hypothetical game concerning two prisoners in a jail who each have the choice of either: snitching on the other (and gaining their own freedom, while the other serves a longer sentence) or cooperating with the other prisoner by both remaining silent (in which case, each prisoner serves a shorter sentence). The game is a paradox—betrayal is the advantageous strategy for each, but when *both* choose the advantageous strategy, it leads to the worst outcome.

The same can be said for split-versus-steal in *Golden Balls*. The "both players steal" outcome is an example of what is known as a *Nash equilibrium*. Nash equilibriums are not necessarily negative outcomes but, rather, eventual outcomes.

Once the players reach this state, no individual player can benefit by altering their strategy. If both players steal, no player stands to gain a thing by changing their strategy and splitting, so in a sense the playing field is balanced. This kind of outcome takes its name from mathematician John Nash, who made great contributions to the field of game theory beginning in the 1950s. Nash was awarded the Nobel Memorial Prize in Economic Sciences in 1994 in recognition of his development of the Nash equilibrium theory, and his brilliance was explored in the Academy Award–winning film *A Beautiful Mind*, which was inspired by the biography of the same title by Sylvia Nasar.

Nash's mathematical genius is legendary, but the film also sought to enlighten movie-goers worldwide about the challenges he faced due to his lifelong struggles with mental illness. It did not, however, reference the bigotry Nash experienced due to persistent accusations that he was gay, despite being married to a woman and having two children. As a young man, Nash was harassed by his classmates for what they deemed to be "homo" behavior and years later, in the summer of 1954, Nash was arrested for allegedly cruising the men's bathroom at Santa Monica beach in one of the local police's regular sting operations cooked up to run homosexual people out of town. Not content to merely arrest him for indecent exposure, the police informed his employer, the RAND Corporation, which then stripped him of his security clearance and fired him.

Nash's career in mathematics was at a high point during the Lavender Scare—when homosexuality was at the center of a nationwide moral panic—which ran parallel to the Red Scare paranoia about communism after World War II. There was enormous pressure to purge the "lavender lads" from government jobs, as homosexuality was actually considered a security risk. Few people were willing to live out-of-the-closet due to the risk of losing their reputations and livelihoods. Gay people, especially those with high-ranking jobs, had to live double lives.

It's important to note that Nash never identified as gay or bisexual and denied the rumors about his sexuality up until his death in 2015. Unfortunately, neither Nash's personal truths nor his public triumphs were enough to protect him from a homophobic society, which starts by indoctrinating children and forces homosexual adults to lie and cover-up their relationships to avoid serious personal and professional consequences. That this issue was ignored by the Hollywood biopic is a subject of controversy; the director and producers have long claimed that his sexuality simply wasn't an important part of his story. But for me, and many others, the decision to completely ignore the prejudice Nash suffered underscores the hostility that LGBTQ people face in the math community and the world at large. It's not the first and far from the last time those who write history have proved very reluctant to associate genius with queerness. In the twenty-first century, we must still hide our truths in order to fit in and be seen as credible and dignified. Navigating a complex maze of choices between honesty and acceptance is a strategic game that LGBTQ people are forced to play from an early age.

Many games are so complicated that reaching the Nash equilibrium may never happen within a player's lifetime. In some games with more than two players, the existence of a Nash equilibrium at all may remain an open question.

Games like *Golden Balls* where Nash equilibriums come into play are interesting for a number of reasons. The simple setup of split-versus-steal reveals some insights into human psychology: when we act in our own self-interest, it can cause us all to be worse off for it. There are real life examples of this phenomenon. Take for example doping in sports. Athletes have been known to use drugs to enhance performance in cycling, tennis, weightlifting, and even ancient chariot racing! If all athletes agree to not dope, the game is fair, but each individual athlete can choose to dope in secret, cheating to gain an

Payoff matrix for cheating at sports

	Your opponent plays fair	**Your opponent cheats**
You play fair	You have a 50% chance of winning	You have a 25% chance of winning
You cheat	You have a 75% chance of winning	You have a 50% chance of winning

advantage. Other athletes are aware of this, and if they think that their competitors will dope, they can feel pressured to dope as well in order to even the playing field. These options can be displayed in a payoff matrix.

The best outcome would be for both you and your opponent to play fair. But maybe your years in training and the weight of your fans' expectations have put you under enormous pressure, so you want to maximize your chances of victory by any means necessary. On reflection, you deduce that cheating is the best strategy. However, if both you and your opponent cheat, the advantages cancel out since the playing field is once again equal—all that's left are the dangers to your health.

Although it may be terrifying to think this way, a nuclear arms race is a similar kind of game. We can all agree that we'd be better off as a planet if no country possessed nuclear weapons. However, this complete disarmament scenario would *not* be a Nash equilibrium. Remember that a Nash equilibrium is an outcome where no players stand to benefit by altering their strategies. But on a planet where no country possesses nuclear weapons, there's an opportunity for some politician to gain an upper hand. The idea is this: if your country is the first to develop nuclear weapons, you'd have the upper hand if any conflict breaks out and your citizens would be safer. On the other hand, if your enemy makes nuclear weapons, then you should also make nuclear weapons to defend yourselves. No matter what the enemy is doing, making nuclear weapons is in your best interest. But if every country follows this logic, *everyone* will have nuclear weapons, and we will again be on an equal playing field just like we were before. But we will all be worse off because now the risks and stakes are much higher!

There's a kind of arms race going on within *Drag Race* as well. Queens must bring 12–15 outfits to wear on runways and in challenges, to harmonize with prompts such as "Best Drag," "Holiday CEO," or "Neon Nights." You can use an outfit that's already in your closet, but most queens buy brand new garments with little to no compensation from the show itself. It's kind of like doping, but instead of performance-enhancing drugs, queens buy performance-enhancing drag.

The choice this time isn't whether or not to cheat, but rather how much money to spend on clothes. If all queens invest the same amount, then they would each be judged on how they choose and style their outfits. However, a queen can gain an advantage by spending some extra coin and upgrading their drag. A custom-fitting corset, a professionally styled wig, and a sparkly earring can make a big difference. Knowing this, if every queen upgrades their drag to the same caliber, they'll all be back on even ground; the judges will have to find other things to rank them on, like dancing prowess or stand-up comedy skills. It may not be illegal, and it may not endanger their health, but doping their wardrobe can put a queen in debt that can take years to pay off. Some queens spend up to $100,000 on preparing for *Drag Race*! (And that number is only rising with inflation.)

What's more, each year the bar is set higher and higher. In early seasons of *Drag Race,* queens could get away with donning outfits they wore to the club, but now there are queens wearing mechanical angel wings, authentic Versace, and LED-light displays worthy of the Met Gala. On *RuPaul's Drag Race All Stars,* expectations escalate even more because some of the queens returning to the runway have become millionaires with connections to designer fashion labels.

There are some seasons of *Drag Race* where the wealth gap between runway outfits is more obvious, and it's heartbreaking to watch talented performers get booted from the competition simply because their clothes looked homemade or were purchased off the rack. Some franchises of the show don't even offer a cash prize to the winner, so the only way for us queens to recoup our investment is through the gigs we can (hopefully) book after appearing. But the hard truth is that not every queen makes their money back. I know more than one queen who regrets appearing on the show because they've struggled to pay rent and settle their debts to designers.

Let me make one thing clear. I don't think that spending money on outfits is as bad as doping, and I don't think that it should be banned! As a viewer, I admit it's exciting to see expensive drag that pushes the boundaries of fashion, probably in a similar way that sports fans want to see athletes working at peak human performance. But as a drag queen, there's a lot of pressure to look glamorous and to constantly impress by wearing brand new outfits. So what can we do to even out the playing field?

I've been told that for some seasons of *Drag Race*, the queens are given a stipend of around $10,000 to help them buy clothes. That's no small budget, but if a queen is still allowed to add their own money to the stipend, it doesn't exactly solve the problem of the wealth gap. An individual queen still has the opportunity to gain an advantage by adding more money to their runway package. Others have suggested that the show place a cap on spending, forcing queens to get creative with what they already have. This makes sense, but I think it would be a hard rule to enforce. Queens who are just starting out in their careers would struggle to obtain new outfits, whereas seasoned queens may already have large wardrobes to choose from. Putting a cap on spending would reward the queens who are able to get clothes for free, either by making outfits themselves, or borrowing from friends or brand sponsors. One of the best challenges on *Drag Race* is the design challenge, where queens are required to walk down the runway in outfits they created while on set, with no help from the outside world. In this challenge, the key advantages are a queen's personal skills in designing and construction; income is less of a deciding factor. We can alter the games we play to make them more fair. Or of course we can decide not to play at all.

One contestant on *Golden Balls,* named Nick, deployed a strategy that shook the rules and ultimately broke the game. During the deliberation stage where most contestants start promising to split with each other, Nick immediately announced he would steal the money, and he told his partner, Ibrahim, that he would later split the money with him after the show, when the money entered his account. Ibrahim tried convincing Nick to change his mind, appealing to his morals, and getting the audience to laugh at him, all to no avail. Nick wouldn't budge and guaranteed he'd steal. This left Ibrahim with no choice. If Ibrahim chose to steal, they would both leave with nothing, but if Ibrahim chose to split, at least Nick would get all the money and Ibrahim might get some of it after the show if Nick kept his word. This could have gone badly if Ibrahim chose to be vindictive and steal out of spite, but Ibrahim ended up choosing to split the money. To Ibrahim, even the small chance of a payout if he trusted Nick was better than a 100% chance of walking away with nothing. To everyone's surprise, Nick also chose to split, so they both won half the pot of money. Nick had found a strategy that no other contestant had thought of before, and he effectively altered the playing field. When the Nash equilibrium points to an undesirable outcome, it's up to us to change the game.

Playing the same game repeatedly teaches players lessons they can use to change their strategy going forward. Vervet monkeys, native to Africa, play a version of *Golden Balls* every time they encounter predators in the wild, such as leopards, eagles, pythons, or baboons. Any individual monkey who spots a predator can choose between sounding an alarm call to warn the troop or hiding in silence. To hide would be the selfish choice, like choosing to "steal" in *Golden Balls,* and is at first glance the optimal strategy. Screaming to let others escape is the altruistic choice, much like choosing to "split." It offers the player no advantage to draw attention to themself, but if everyone works together, all the monkeys win. As it turns out, vervet monkeys are taught to make almost 30 different alarm sounds from infancy. It appears that over the long run, cooperation works better. If you are a selfish monkey who chooses not to warn others, you later may find yourself with your back turned at the wrong time. If you live to tell the tale, you will learn that screaming an alarm call is better for the troop, and hopefully one day someone else will risk their life so that you can escape with yours.

The good news is that humans can be altruistic, too. The thing about game theory is that most game theorists operate on the assumption that humans are perfectly rational logic-machines who always act in their own best interest. I think that's an awfully big assumption! In real life games, we don't always go for the selfish choice, and if experience tells us anything, it's that humans are anything *but* rational.

We have this thing we call "sentimental value." When my dad died, I inherited his Rolex, which I've only worn once—on my wedding day. The watch is probably worth over $10,000, but I'll never sell it. (Unless RuPaul herself asks me to go on All Stars, then maybe I'll need some Louboutins!) Now I, myself, would never dream of spending $10,000 on a Rolex even if I found the exact same model at a store. The thought of spending that much on a watch is absolutely insane to me! To be honest, I don't even find the watch that cute. Personally I prefer the look of a simple watch over one with lots of knobs and doohickeys. But to a perfectly rational decision-maker, this behavior doesn't make any sense. On one hand, I believe the watch I inherited from my dad is essentially priceless; it's worth far more to me than the $10,000 it could fetch at a pawnshop. But on the other hand, if that exact same watch were sitting in a pawnshop with a $10,000 price tag, I would say it was totally overpriced. This behavior may not be rational, but I bet it would make perfect sense to anyone who owns something with sentimental value, whether it's a family heirloom or your childhood Pokémon cards. This bias is known to psychologists as the *endowment effect*.

Even a game like rock-paper-scissors is subject to psychological bias. According to game theorists, the best strategy to win at rock-paper-scissors is a mixed one—play all three hand positions randomly to stop your opponent from exploiting any patterns. In real life, people can be predictable. Most people prefer to play rock as their first choice, presumably because it feels safe and sturdy. When researchers in China paired up 360 students to play 300 rounds of rock-paper-scissors, they found that rock was played 36% of the time! They also found that players like to repeat a winning strategy. If someone wins by playing rock, they're much more likely to play rock again rather than choosing a different hand position. Game theorists call this the *win–stay, lose–shift strategy*.

When accounting for human emotion, strictly logic-based strategies don't cut it. And marketers are aware of this. I recently watched a video online of someone confessing to spending over $800 a year for a gym membership they don't use simply because they were afraid of confronting the aggressive sales reps at the gym, who charge a $40 cancellation fee. (I'd be too scared to cancel, too.) And while some of us overpay to avoid confrontation, others overpay for the flash and status of designer names and recognizable logos. I once paid $500 for a vintage orange Gianni Versace jacket because it matched an outfit I was putting together. It looked like a basic cropped blazer you could buy at any thrift store, but I couldn't resist the sound of "vintage" and "Versace" put together. I only wore it once!

Then again, we might also overpay to support a small business that uses sustainable packaging, or kick in a little extra to help a struggling server who makes a living off of their tips. We might shovel the snow off a neighbor's driveway even though we're busy or stand up to a bully even if it means putting our own safety at risk. Perhaps we are simply like monkeys who put ourselves at a short-term loss in order to aid the common good, in hopes that someone else will be as selfless when we're down on our luck. Then again, maybe we're just irrational.

Behavioral Dragonomics

If game theory is the study of rational decision-makers, then the study of irrational decision-makers is called *behavioral economics.* Behavioral economists study the effects of emotions, culture, marketing, and psychology on human decision-making. As much as we try to resist, our emotions get in the way of our strategic thinking. I promise you that is true of everyone, from athletes and real-estate moguls to mathematically minded drag queens.

As a hard-core *Drag Race* superfan and a very competitive person, I never would have dreamed in a thousand years that I'd make it onto the show and then lose my desire to win. But I can admit here that my emotions got the better of me.

Drag Race is a complex game because the rules aren't as precisely defined as they are in rock-paper-scissors or in chess. The goal is to impress a panel of judges who rank you based on their personal tastes, and it was tough for me to figure out exactly what they liked and what they didn't without an established algorithm. So in classic analytical fashion, I formed my strategy based on what I learned from carefully watching previous seasons of the American show. I figured that my strongest asset was my confidence, and I'd use it to shroud any self-doubt that could only trip me up.

I entered the "werk room" with a big bra and an even bigger ego. Our first challenge was a photo shoot on top of a "wintery mountain," with a gigantic wind machine blowing our fake eyelashes to and fro. The judges sat watching, and cracked a few jokes about how I looked like Ongina or any other Asian *Drag Race* queen. Making little quips and jokes during the challenges was the judges' job. After we descended the mountain, it was revealed that I'd won the mini-challenge! I was off to a great start.

The episode's main challenge was to sew an outfit that showcased our signature style by using materials from a themed box filled with fabrics and accessories. Since I won the mini-challenge, I got to assign a box to each queen. Obviously, I gave myself what I thought was the best box, one enticingly covered in gold. Right beside me stood a queen named Jimbo, who whispered that he wanted the rainbow box, so I agreed, like the altruistic vervet monkey I am. From there, I went down the line assigning the boxes pretty randomly. I didn't need to sabotage anyone with my box choices because I was confident enough in my own ability to slay the challenge. After all, I knew my way around a sewing machine and many of my costumes were my own creations, so I was sure I had this one in the bag. Thus empowered, I decided to readjust my focus toward the "game within the game"—winning over the producers by cranking up the entertainment factor. I spent the afternoon prancing around the room boasting about my mini-challenge victory and putting on a RuPaul persona to interview the other queens. That didn't go terribly well, but I was undaunted.

When I settled down to the actual task, I sewed a gold jumpsuit with a glittery collar and lapel—a signature Kyne outfit and one I was used to making. In addition to the material and supplies in the box, the producers had a shelf of basic sewing supplies we could use; these included a store-bought corset and a hoop skirt. Hoping to win some extra points for creativity, I took apart the hoop skirt and inserted the plastic boning into the pant legs of my jumpsuit to make them end in big bell-bottom rings. Then I hot-glued on some Christmas ornaments—golden balls, no less! But as soon as I tried my creation on, I realized I had a problem. The ornaments were falling off with every step I took! I immediately began berating myself. Why didn't I try it on sooner? Why did I waste so much time frolicking around bragging about how I was going to win?

It was too late for me to change things, so the only solution was to waddle down the runway with my feet apart like a gay penguin rushing to the bathroom, golden baubles scattering merrily in my wake. I told myself, "Kyne, you can get through this. Even if your outfit is falling apart, it'll be okay if you can make the judges laugh." So I joked that the balls were meant to fall off because, "Everybody loves balls!" They were not amused.

At that point I was completely shaken up and embarrassed, and frustrated that the judging seemed to be unfairly harsh. I lashed out against a queen named BOA because I was totally jealous that the judges fell in love with her stage presence and charisma, despite her wearing potatoes and green ribbons glued to a corset. She made the judges laugh without even trying, something I could never do! I was a terrible sport, but I didn't see it that way at the time. I went back to my hotel room that night and cried hysterically, thinking how much of a fool I had made of myself.

Thankfully, I survived the first episode. But, judging by my reaction, you would have thought that I was the queen who was sent home!

However, I saw the next episode as a chance to rewrite the script and refresh the game with a brand-new, even if still massively complicated, payoff matrix. I vowed to take the lessons I learned from the first game into the next one. My new strategy looked like this: don't take yourself too seriously; make the judges laugh; and be humble.

Our second episode involved an acting challenge, where we were split up into teams and directed to memorize a script. Our main judge, Brooke Lynn Hytes, entered the werk room and promptly came over to me and asked how I felt about almost being eliminated the previous week. I tried to sarcastically shrug it off and change the topic. After all, this was a new week! "Me? In the bottom? You must be confusing me with someone else!" But she wasn't deterred. So I archly informed her that I forgave her for putting me there. Yikes . . . another joke that didn't land very well! Moving right along, I apologized to BOA for starting a fight, and we swiftly made up. She picked me to be on her team, and I felt I nailed the second challenge. I knew all my lines and managed to get a thumbs up from a lot of the people in the room. Even the tough judge, Jeffrey Bowyer-Chapman, told me I did a great job. You can imagine my shock when the next day on the runway he told me my group had completely overshadowed me! I offered no more comebacks or jokes; I had learned my lesson. The judges handed down my sentence: I would be lip-syncing for my life.

This is usually the moment where I would be shouting at the TV screen, cheering on the embattled queens, telling them to fight until the bitter end. But everything changed when I was standing in their pumps.

As I waited for the crew to set up the song, I felt exhausted, confused, and demoralized. I felt as though no matter how hard I tried, the judges just didn't like me. My best wasn't good enough. I remember looking up at the ceiling so that the tears would stay put in my eyes and not ruin my makeup, thinking about my husband at home rooting for me. I thought, even if I lose this competition, I'll still have him.

God, I was such a drama queen!

I ending up losing the lip sync to a queen I adore, Tynomi Banks. As I sashayed away, I felt a wave of emotions: at first relief, then shame and disappointment, and then relief again. I was glad I didn't have to fight anymore. I sobbed hysterically and could barely finish delivering my goodbye message to the camera before the producers ushered me away to wipe off my makeup and pack up my clothes.

I lost the game fair and square.

At first it stung, but I made my peace with it. A year later, I relived those emotions when my season began airing on TV, and everyone saw it for the first time. I started receiving messages from people telling me that they hated my attitude and that I needed a good slap in the face. After a few weeks passed and those fans moved on, new viewers would start streaming the show, watching it from the beginning, so I'd receive more hate mail. I had to disable comments on my social media accounts because these messages would ruin my entire day. I started to associate so many negative memories with what was once my favorite show that it became difficult to watch anymore. To be completely real with you, my biggest flaw (and insecurity) is that I always think about how others perceive me. I know I'm not perfect, but I'll be damned if you remind me of it!

Despite everything, I don't have any regrets about my time on *Drag Race.* If I had done anything differently, I may not have learned as much as I did. I discovered that I can be stubborn, that I can be overly confident in my abilities at times, and that I had a lot more to learn than I thought. Most importantly, losing motivated me to play my own games, ones where I can write the rules and decide for myself what it means to be a winner.

Drag Race taught me a few major lessons about the limitations of game theory, the role of behavioral economics, and most importantly, myself. I'd thought that all 12 of us were there fighting to cut the largest share of a cake, but I've realized that wasn't the game at all. Some queens like their slice to be topped with the most showstopping outfit, others covet the sweet frosting of being the funniest queen in the room, and then there might even be an occasional queen who prefers pi.

After the season ends, queens pursue different ventures, from acting, hosting, or singing, to stand-up comedy or fashion shows. Some queens keep drag as a hobby, and others make it a profession. It's possible to walk away a winner with only a slim slice of the cake if it's also your favorite flavor. True victory is contentment.

Most games in life don't end with crowns, cash prizes, or sashaying away, which is a good thing because life is full of games and that would make for too much drama every day. No matter how high or low the stakes, game theory offers unique mathematical insights into the inner workings of rational decision-making and how best to strategize in order to maximize your payoff. But think carefully about which payoff you intend to maximize. Do you want to win by having the most money, or the most fun? Would you rather get to your destination as fast as possible, or take the most scenic route? Do you want the largest slice of cake, or does it matter more that your friends are happy with their share? The best game strategies are highly dependent on how you personally measure success. Winning isn't everything. Sometimes it's better to just enjoy being a player.

CHAPTER 4

Luck Be a Ladyboy

You can do whatever you want in life. You are free to ditch school and join a traveling troupe of drag queens, or quit your job and become a cowboy, or jump in a car and rob a bank. Life is all a matter of gambles. I took a big gamble when I joined *Drag Race*, and despite my strategizing, I still lost! But the thing no one prepared me for was the competition that began after *Drag Race* ended. Being part of the *Drag Race* cinematic universe meant that we contestants were sort of like collectible trading cards—every fan had their favorites, and we were categorized into fashion queens, social media queens, comedy queens, heroes, and villains. Who had the most fans? Who had the most expensive outfits? Who had the best merchandise? It started to mess with my head.

So I decided to try something new: posting math riddles on TikTok, an app where nobody knew who I was. Back then, math wasn't a part of my drag act; I was living a secret double life, like Hannah Montana. By day, I was attending university lectures in pure mathematics, and by night I was doing the splits in some bar and making YouTube videos about styling wigs.

Whenever I mentioned my math degree to any of my drag friends, they would either tease me for being a nerd or stare at me in confusion. I certainly never thought that anyone would be interested in hearing me talk about math while dressed in drag, until I made my first video on TikTok. It was meant to be a joke, and my being dressed in an elaborate wig and makeup while discussing fractions was part of the absurdity of it all. But instead of the nonresponse I was expecting, I got loads of comments from people telling me that my videos were entertaining and inspiring.

The videos went viral, and before I knew it, I was getting interviewed by Buzzfeed and CNBC. I even got a shoutout from Canada's prime minister!

As it turns out, I didn't need a TV network to find an appreciative audience. But being on *Drag Race* did help, and for that I'll always be grateful. The big break I always dreamed of ultimately wasn't meant for me; instead, the silly thing I thought would never work ended up changing my life.

Actually, a lot of my victories have started out looking to others like absurdities. When I first started wearing makeup to school, I had friends pull me aside and say, "Kyne, you know makeup is only for girls, right?" When I started my YouTube channel, people teased me about that too. And when I began teaching math in drag, that idea was met with a lot of skepticism and ridicule.

One Christmas Eve, I received a message from a stranger who had just watched one of my YouTube videos. He told me he thought we had a lot in common and that he'd like to take me out on a date. The only problem: he lived in the United Kingdom and I was an ocean away in Canada. I told him I had never been to Europe and that I didn't have much interest in visiting. I was playing a little hard to get, what can I say? But we continued to exchange messages. I told him about my family, my past, and my dreams for the future. Messages turned into phone calls, which turned into video connections. In the beginning of January, I booked a flight to meet up with him in person. It was the most spontaneous thing I had ever done, and everybody told me I was crazy. But I calculated that the risks of him being a kidnapper were pretty low. In my mind, the only risk was that in person he would be different from the boy I liked over the phone. If the chemistry wasn't the same face-to-face, I figured I would at least go back home with a funny story to tell. Instead, I fell head over heels in love.

Reader, I married him! *Disclaimer: your results may vary.*

We have the YouTube algorithm to thank for our happiness. What were the chances of us meeting under any other circumstance? I don't believe in fate, but I do believe in the beauty of randomness. And there's always a little randomness, even in the things you think are certain.

Every decision we make rests in part on a subconscious cost-benefit analysis of our different options. We might predict what the other players in our world will do, but many of our decisions are based on rough bets and estimates. These decisions can be hugely important or seemingly inconsequential, and sometimes we don't know which is which. Should I do what my parents want me to do, or should I try something fun and dangerous? Should I wear blond hair or black? Should I reply to this DM? If I go out looking like a slob, will I run into anyone I know?

Our lives are shaped by our decisions, and our decisions are shaped by uncertainty. There is no running away from Lady Luck. . . . Or is there? What if we could conquer randomness by using math and science to predict the future?

If you had access to a GPS database that tracked everyone you knew, you'd be informed beforehand about whether you would run into someone important while you were out, and you could decide, based on that, how you want to get dressed, or whether you want to go out at all. If you had full schematics and building information about a bank, and even access to the security system, you might be able to rob it and get away with it. And if I could count cards and had the memory of a computer, I could take home millions at the casino without relying on my fickle friend, Ladyboy Luck, at all.

We might try to define randomness as merely the absence of information; perhaps everything we perceive as random is actually completely rational and predetermined, and we are just sadly uninformed. People once thought that eclipses and comets were unpredictable omens from the gods—signs to crown a king, to found a city, to stop a war, or to start one. Now we know exactly how they work and can even accurately predict the appearance of some of them hundreds of years in advance. What was once seen as random was simply a natural, predetermined routine running its course. So why do people cross their fingers for blue skies on their wedding day or knock on wood for snow on Christmas, even though the weather is as deterministic as the orbit of planets? Maybe we just have a funny relationship with randomness.

Years ago, Spotify users complained that the shuffle button wasn't random enough. Their shuffled playlists felt like a deck of cards with a bunch of spades clumped together. But Spotify playlists aren't shuffled by hand like a deck of cards; they're shuffled by a prewritten algorithm. When developers looked into it, the problem was that the shuffle button was *too* random. The original algorithm generated perfectly random shuffles, which meant that every possible ordering of songs was equally likely. But often this created clusters of, say, four Britney Spears songs back-to-back, followed by three Spice Girls songs. Users didn't like this, so Spotify changed the algorithm to spread apart songs by the same artist, making the playlist feel just random enough. Which, of course, means it isn't completely random.

To our minds, randomness ought to mean shuffling things until they become evenly spaced, but true randomness naturally creates clusters: sometimes you flip a coin and get five heads in a row, or a roulette wheel lands on black six spins in a row. In a famous game of roulette at the Monte Carlo Casino in 1913, the ball fell on black 26 times in a row—a 1 in 66 million chance! You don't have to be a mathematician to know that this was unlikely. But gamblers made a crucial mistake by betting that the streak would end. They lost millions of francs betting on red, with the mistaken idea that red was somehow now *more* likely than black because of this long streak. Instead, every individual spin of the wheel is independent from any other spin; meaning that, in a truly random scenario, the past has no influence on the future. A streak of 26 blacks in a row may indeed happen once in 66 million games, but the probability of landing on black on the next spin is still the same as the probability of landing on red, no matter what happened in the past.

The same bias appears in couples who have four boys in a row and so they bet that their next baby is likely to be a girl. There's a notion that the law of averages mandates that the future must balance out the past, and a long streak of boys must be balanced out by a streak of girls. But this isn't a real mathematical law; it's a cognitive bias known as the *gambler's fallacy.*

Millennia ago, people threw bones and made decisions based on the way they fell, believing that the gods guided the bones. Nowadays, people might flip a coin to decide whether to eat Chinese food or Mexican food for dinner. Leave it up to the gods to decide! In some sense, randomness equates to fairness. At the same time, randomness also represents the chaos, frenzy, and unfairness of nature. Randomness means that some food left behind in a park made your dog sick, a flat tire made you late for a job interview, or a bat in Wuhan killed millions of people around the world, ravaged the global economy, and postponed your wedding to the boy you met over the internet on Christmas Eve. (Just for instance.) The random act of simply finding yourself in the wrong place at the wrong time can change the course of your life catastrophically. Or it can bring you together with the love of your life.

Slaying Uncertainty

In sixteenth-century Italy, a Franciscan monk named Luca Pacioli started to toy with quantifying randomness using the thought experiment of an unfinished game. Imagine you are playing rock-paper-scissors with a friend and you agree the winner of 2 out of 3 rounds gets $100. Perhaps you remember from the previous chapter that 36% of players choose rock, so you try to exploit this and play paper. But in the first round your opponent is one step ahead, and has chosen to play scissors! The score is 0–1. You're losing, but you still have a chance to settle the score in the next 2 rounds. Before you can make your next move, the game is abruptly interrupted; maybe someone pulled the fire alarm or your mom is calling you for dinner. If you knew you would be unable to resume your game later and were forced to split the $100 now, how should it be split? Should you split it 50–50 since the winner hasn't been decided yet, or should your friend be rewarded for having an advantage?

It's like trying to bet on who will win a season of *RuPaul's Drag Race* based only on the first episode. If you had to bet on somebody, all signs point to the winner of the first challenge. But how much support should we give the underdog? How common is an upset? What Pacioli and many others were on the verge of discovering was the branch of mathematics we now call probability theory.

If you handed this problem to a student taking a high school math class, they might approach it like this. In order for you to win 2 out of 3 rounds of rock-paper-scissors after losing the first, you will have to win both of the next 2 rounds. Let's say you have a 50% chance of winning round 2, and a 50% chance of winning round 3, which multiply together to make a 25% chance of winning both rounds. Thus your friend has a 75% chance of winning best 2 out of 3, given that they've already won round 1.

Another way to visualize this is by listing all your possible outcomes. The first round was a loss for you, so we can only extrapolate what could happen in the next 2 rounds:

loss, loss, win
loss, loss, loss
loss, win, win
loss, win, loss

Only one scenario (in bold) results in a 2 out of 3 win for you, so you only win in 1 out of 4 outcomes, or 25%. Thus the fairest way to split the money is giving $75 to your friend and keeping $25 for yourself.

This kind of analysis was completely novel in the sixteenth century, at least in the West, where mathematicians were only beginning to think about probability using numbers. Dice games have been common for millennia, but the mathematicians of ancient Greece and Rome didn't seem to think about assigning numbers to random chances. If your son was sent off to fight in a war, you might say that he's either likely to return home or likely to die in war, depending on his age, health, and what you know of the ferocity of the opposing side. But it might not occur to you to represent your degree of hope using a number between 0 and 1. The outcome of war and the fall of the dice were both as fickle and unpredictable as the gods.

Whenever I talk about probability online, I inevitably get comments about the hot take that every probability is 50% because every event—no matter what it is—either happens or doesn't. A soldier has a 50% probability of returning home from war because either they do or don't return. Likewise, your lottery ticket can only either be a winner or a loser, your plane either lands safely or it doesn't, and you either win *Drag Race* or you don't. These are all binary outcomes, so you'd be forgiven if you think that the chances of them happening are 1 in 2, or 50%.

The thing is, if we use probabilities other than 50%, our predictive ability becomes much more powerful and much more accurate. Probability theory gives us a way of controlling and understanding randomness to mitigate risks and minimize harm. It's no wonder that probability theory developed during the Renaissance, a time when many people started believing that we humans, not the gods, are the agents of our own fate. We seek to triumph over unpredictability. Whether or not you believe that the gods (or God with a capital *G*) are spinning the threads of fate, there's no doubt that human decisions can affect the future. So shouldn't we be able to predict at least some things about it?

Well, most things in life aren't as predictable as eclipses or comets. It might be easy to predict the outcome of a coin toss if you drop it from a height of 1 cm. You could predict how many times it will spin before hitting the ground, maybe even from a height of 30 cm. But what if you drop it from Toronto's CN Tower? Even if you knew the coin's weight, proportions, the initial velocity and angle of the toss, and the atmospheric conditions—is that enough to know for certain how it will fall? Have you accounted for the probability that it will smash into a pigeon on its way down? Did you factor in the cheering of the sports fans in the Scotiabank Arena nearby? There's just no knowing for sure. Somewhere a drag queen doing a backflip in an underground club sets off a tsunami on the other side of the world, or a little kid buying a chocolate bar from a convenience store causes the stock market to crash.

A GPS tracker that tells us when and where we're going to run into our one true love doesn't exist. We simply don't have access to all the information we think we need; meaning we can never know the future for certain. We live our lives through rough estimates, rules of thumb, and gut feelings. If a carton of milk smells weird, we don't take it to a laboratory to test whether it will give us food poisoning; we just toss it out. If your friend had a few glasses of wine hours ago but they still seem pretty tipsy, you don't pull out a breathalyzer to test their blood alcohol level before letting them drive; you just call them a taxi.

Sometimes our intuition fails us and leads to disastrous situations. But total omniscience is impossible. Luckily there's a sweet spot in the middle, and that's where we find the branch of math called *probability*.

Probability seeks to quantify how *likely* an outcome is on a scale of 0 to 1, from 0% certainty to 100%. Most things fall somewhere in the middle. There will always be an element of uncertainty, a small agent of chaos that makes life thrilling and adventurous. It's Ladyboy Luck who gives us freedom.

Probability replaces the need for prophets and oracles. Studying randomness recognizes that we are free agents who decide our own fate. Want to take the path with the least risk, or the one with the greatest reward? We can choose for ourselves the amount of risk we want to take. And with good mathematical models, the uncertainty can be made smaller and smaller, allowing us to take larger, more calculated risks. Probability gives us the best chance, without magic powers, at predicting the future, and it can be so powerful that it may as well *be* a magic power. But predicting the future isn't easy. Studying probability comes with a wealth of perplexing paradoxes, which we'll navigate together!

Now is a good time to finally define probability. What does a number between 0 and 1 really mean anyway? We start with the following definition:

$$\text{Probability of an event happening} = \frac{\text{Number of favorable outcomes}}{\text{Number of possible outcomes}}$$

If you're auditioning for a season of *Drag Race*, which is casting 12 queens, the probability that you'll make it onto the show is theoretically 12 divided by the number of applicants, which can be anywhere from dozens to hundreds or thousands. Suppose a season gets 3,000 applications: only 0.4% of queens who apply will successfully make it onto the show. That probability improves, though, because among all who apply, only about 1,000 will be invited to submit full audition tapes, which is a lengthy process that involves showing off your best celebrity impersonations, your most fabulous costumes, and your out-of-drag personality. If you make it to the audition tape review, you'll have about a 10% chance at doing a phone interview with a producer, as they'll pick about 100 of their favorite audition tapes. Once you're in the top 100, you'll have a 12/100 or 12% chance at being cast!

The producers will then narrow the group down to 20, who will be psychologically and physically evaluated by a doctor, but if you make it to the top 20, you'll have a 12/20 or 60% chance at making it onto the show. If you're lucky enough to be a competitor, you might calculate your chances of winning as 1 in 12 (8.3%), if there are 12 competitors in total. If you've made it to the top 2, your chances of winning go up to 50%! There's a big difference between an 8.3% and a 50% chance of winning.

Sheer Cold, Bad Bets, and Hidden Goats

If you've ever played Pokémon, you already have some first-hand experience at probability and randomness, which is part of what makes the game exciting (in addition to the cool graphics). In any patch of grass, you may encounter a Pokémon you already have, but there's also a chance that you'll encounter one you've never seen before. Winning a Pokémon battle takes more than just skill and training; it also takes luck. There's a small chance that your attack will miss its target, but there's also a chance it lands a critical hit and deals extra damage, which can make all the difference when battling gym leaders and champions.

You can teach your Pokémon specialized attacks, like Aerial Ace or Swift, that have 100% accuracy and never miss, or you can teach it attacks that have 70% accuracy, like Blizzard or Thunder. In exchange for the lower accuracy, Blizzard and Thunder are far more powerful and deal more damage than Aerial Ace and Swift. There are also moves like Sheer Cold that only have 30% accuracy but result in a one-hit KO if they land. As a Pokémon trainer, it's up to you to decide whether to use stronger moves that have a high likelihood of missing or weaker moves that are more reliable.

We can compare attacks mathematically using what we call an *expected value function.*

All an expected value function does is add up all the possible outcomes, but only after we place different weights on them based on how likely they are to occur. Consider the move Blizzard, which deals 110 points of damage with 70% accuracy. On average, we can expect it to deal $110 \times 0.70 = 77$ points of damage. Aerial Ace deals 60 points of damage with 100% accuracy, so its expected value is $60 \times 1.0 = 60$ points. On average, Blizzard does better, but it's not as good as Jump Kick, which deals 100 points of damage with 95% accuracy, giving it an expected value of $100 \times 0.95 = 95$ points.

In the real world, we are often faced with high-risk, high-reward decisions, which can feel like gambling. Playing the lottery is a bit like using the Sheer Cold attack—you're betting on a very small chance that the game can change drastically in your favor. Just as we used the expected value function to assess Pokémon moves, we can also use it to compare lotteries, and decide whether they're even worth the cost of the ticket. For this calculation, we need two pieces of information: the probability of winning and the potential reward. The jackpots are usually plastered in big numbers at the top of every lottery advertisement, but the probabilities of winning them often take a little more investigation to uncover.

Riddle: The popular lottery game where I live is Ontario's Lotto 6/49, where you choose 6 numbers from 1 to 49 and can win up to $20 million. What's the probability of choosing the correct 6 numbers?

Solution: All we have to do is count the number of favorable outcomes and divide by the number of possible outcomes. Everyone's favorable outcome, of course, would be to win by choosing the correct 6 numbers. But there's not only one way to choose the correct 6 numbers because you're allowed to write them in any order, and the numbers don't have to be ordered from least to greatest.

So how many ways can we shuffle around the same 6 numbers? If this question sounds daunting, let's start small. How many ways can we shuffle around 2 numbers? Suppose the 2 numbers are *A* and *B*. We could write them with *A* first and *B* second, or vice versa. There are only 2 ways to shuffle 2 numbers. How about 3 numbers: *A*, *B*, and *C*? There are 6 possible shuffles: *ABC*, *ACB*, *BAC*, *BCA*, *CAB*, *CBA*. Notice that 6 is 3×2, which is also the same as $3 \times 2 \times 1$.

How about shuffling around 4 numbers, *A*, *B*, *C*, and *D*? This list will be longer, but we can make use of our previous work to save some time. If we want to put *A* first, there are 6 ways to shuffle *B*, *C*, and *D* after it: *ABCD*, *ABDC*, *ACBD*, *ACDB*, *ADBC*, *ADCB*. If instead we want to put *B* first, there are 6 ways to shuffle *A*, *C*, and

D after it: *BACD, BADC, BCAD, BCDA, BDAC, BDCA*. There are also 6 ways to put *C* first and another 6 ways to put *D* first, so there are a total of 4 times 6, or 24 ways to shuffle around 4 numbers. Notice that 24 is 4×6, which is $4 \times 3 \times 2 \times 1$. The pattern will continue. The number of ways to shuffle 5 numbers will be 5×24, or $5 \times 4 \times 3 \times 2 \times 1$. Finally, the number of ways to shuffle 6 numbers is going to be $6 \times 5 \times 4 \times 3 \times 2 \times 1$. Another way to interpret this: there are 6 possible numbers you can nominate for the first spot, 5 leftover possibilities for the second spot, 4 for the third spot, 3 for the fourth spot, 2 for the fifth spot, and by the sixth spot there's only 1 number left: $6 \times 5 \times 4 \times 3 \times 2 \times 1$ is 720. We have a name for this mathematical calculation, and it's called a factorial! It's denoted with the exclamation point!

$$6! = 6 \times 5 \times 4 \times 3 \times 2 \times 1 = 720$$

So there are 720 possible orders in which you can choose the same 6 winning numbers, and that's our number of "favorable outcomes." How many "possible outcomes" are there? To find out, we can use a similar strategy. Out of all the 49 numbers you can choose, you can use any of the 49 in your first spot. For the second spot, you have to choose among the 48 numbers you didn't choose first, since you can't repeat choices. You then have 47 choices for the third spot, 46 choices for slot 4, 45 options for slot 5, and 44 options for slot 6. If we multiply these together, we get:

$$49 \times 48 \times 47 \times 46 \times 45 \times 44 = 10{,}068{,}347{,}520$$

So there are 720 favorable outcomes out of a possible 10,068,347,520. This means the probability of a single ticket winning this lottery is 720 / 10,068,347,520 which is 1 in 13,983,816, or 0.000007151%.

Now that we have the two pieces of information we need, we can multiply them together to find the expected value of this lottery:

$$\$20{,}000{,}000 \times 0.00000007151 = \$1.43$$

Not a bad bet—the only problem is that the lottery tickets cost upwards of \$3. So you can expect on average to lose $1.43 - 3 = \$1.57$ each time you play. Or, put differently, if you played the lottery enough times to actually win, then on average, you'll have spent more on tickets than what you'll recoup in winnings. Most lotteries work this way. It's the only way the lottery owners can make a profit.

But the house doesn't always win. In 2003, a retired couple Jerry and Marge Selbee noticed a mathematical flaw in the Michigan state lottery. Jerry calculated that on certain days when the jackpot rolled over from a previous week of having no winner, the expected value of the game was greater than the ticket price. *Cha-Ching*! That doesn't necessarily mean that every ticket is going to be a winner, but it means that if you buy many tickets, then on average your wins will overcompensate your losses. So the Selbees pooled together thousands of dollars, and won thousands more. In 2012, the Michigan lottery closed the loophole, but the Selbees found a similar lottery happening in Massachusetts, so they drove 900 miles to buy hundreds of thousands of dollars' worth of lottery tickets. A group of math students at MIT also caught on, and they took home millions of dollars before the state realized what was going on and phased out the game.

The Selbees and the MIT students took advantage of a mathematical theorem called the *law of large numbers*. If a game is repeated a large number of times, then the average payoff will get closer and closer to the mathematical expected value. This is the same exact mathematical phenomenon that casinos use to secure their profits. The odds of winning any game in a casino is somewhere below 50%. Experts say that blackjack has the best odds of any game—you have a 49% chance of winning. If you're lucky enough to win something, you should take the money and run. The longer you play, and the more people who play alongside you, the more the casino wins. The 51 licensed casinos on the Las Vegas strip generate $7 billion of revenue annually from games and slot machines (more than the annual gross domestic product of some small countries).

Americans spend nearly $100 billion on lotteries each year, most of which are run by state governments. Interestingly, US law prohibits anyone from mailing, circulating, or advertising any sort of lotteries, with the notable exception of lotteries run by state governments. Some of the resulting revenue is put to good use. After paying out prizes, administrative fees, and advertising expenses, the Pennsylvania lottery dedicates the remainder of its proceeds to programs for the elderly, while the lottery in Georgia forms the sole source of funding for the state's HOPE Scholarship, which has helped more than 1.8 million students attend college. It seems almost like a win for everybody involved—at a small cost to the players, a good cause receives funding and one lucky winner receives a big jackpot. Since many states use lotteries to balance their budgets, they're incentivized to sell as many tickets as possible, and that means advertising to as many people as possible.

State-run lotteries, like the Massachusetts Cash WinFall, are even exempt from truth-in-advertising regulations; meaning, they can advertise as aggressively as they want. There's something a bit troubling about a government promoting an addictive habit of frivolous spending, knowing that the math is working against the average player, especially when you consider that poorer people are spending a larger portion of their income on lottery tickets than wealthier people are.

Personally, I'm not a big fan of lotteries or casino games. I think gambling is a waste of money. In my opinion, the only reason to do it is if you're in it strictly for giggles. I've only once gambled on a machine; I was on a vacation with friends, we each spent $10 or $20 and won nothing—but we had a good night! What I do love, though, is the math behind the games. My favorite games of chance are the ones where the probabilities are totally counterintuitive. They're more surprising, more ridiculous, and more fun! Let's take for example the simple game of flipping coins.

> **Riddle:** Suppose I flip two coins at the same time. Without revealing them to you, I look at them both and tell you that at least one of the coins landed with heads facing up. What is the probability that the other coin also landed with heads facing up?
>
> **Solution:** Your first thought might be to say 50%. After all, if one coin was a heads, then the probability that the other coin was also a heads is 50%. It's not like one coin can influence the other, right? And it's true—coin flips *don't* influence each other. You can flip heads 30 times in a row, and the probability of getting heads on your 31st flip is still 50%. But in this scenario, the math is very surprising. The probability of getting heads on your other coin is actually approximately 33%! Here's why. There are four possible outcomes when flipping two coins: HH, HT, TH, or TT. However, since I, the dealer, revealed that at least one of the coin flips resulted in heads, we can eliminate TT from the calculations. That means there are only three possible outcomes: HH, HT, or TH. Only one

of these outcomes (HH) is favorable, since we need both coins to land heads up. We have one favorable outcome out of a possible three; thus the probability we're looking for is 1 in 3. If you're still not convinced, think of it this way: when flipping two coins, in general, you're equally likely to get two of the same face (HH or TT) as you are to get two different faces (HT or TH). However, in the hypothetical scenario where I tell you that at least one of the coins landed with heads facing up, I've completely eliminated the possibility of getting TT; meaning, the odds have shifted in favor of the coins landing on two different faces, so the chances that the second coin also landed heads is closer to 33% instead of 50%.

A few years ago I shared a similar version of this riddle in a video on TikTok and it went viral, splitting my entire audience in half, arguing heatedly over the answer. Someone much more famous than me (who actually created one of my favorite childhood TV shows) reposted my video and told everyone I had it wrong. He kept on insisting that I was overcomplicating the question, and we had a little math feud, which spanned, like, five videos. Days later, he finally admitted he had it wrong. He was eventually convinced after seeing a computer simulation of the riddle. You can program a computer to digitally flip a pair of coins thousands of times, and then eliminate all the instances where TT appears, to ensure that at least one of the coin flips landed with heads facing up. Among those remaining coin flips, you can confirm that about 33% of them are HH and 67% are HT or TH. His surrender was a moment of sweet victory for me, but mostly I was just busy fangirling.

Perhaps the most famously controversial probability paradox is the Monty Hall problem, named after the host of the old game show *Let's Make a Deal*.

Riddle: Imagine you're a contestant who is given the choice of one of three doors, and you get to keep the prize behind whichever door you choose. Behind one of the doors is a necklace featuring the 128-carat Tiffany Yellow Diamond, valued at $30 million and famously worn by only four women since its discovery in the 1800s, including Audrey Hepburn, Lady Gaga, and Beyoncé. Behind the other two doors are goats. (I know goats are majestic and beautiful and we'd all love to have one, but for the sake of argument, just view them as duds that no one wants.) The three doors all look the same to you, so you choose one at random, but before opening it, Monty opens one of the doors you *didn't* choose and reveals a goat behind it. Then, he asks if you would like to switch choices to the only other unopened door. Should you switch or stay?

Solution: At first glance, it should seem that switching makes no difference at all. Monty revealing one goat doesn't change what was behind the other two doors, so why should the goat-reveal affect your decision?

There are two doors left, and behind one of them is the diamond and behind the other is a goat, so you should have a 50% chance of choosing either prize at either door, right? In 1990, Marilyn vos Savant put this problem on the map when she wrote about it in her *Parade* magazine column, *Ask Marilyn*. She wrote that switching doors is always better, and she was correct. Think about it this way: after you make your initial choice, Monty *always* reveals a goat behind a different door, leaving the necklace and the other goat as the only remaining options. Switching will *always* result in getting the opposite of what was behind your initial door!

If your initial choice was a goat, then you will switch to the diamond, and if your initial choice happened to be the diamond, then you will switch to a goat. If we do switch, we would obviously like to switch to the dazzling necklace. So switching is advantageous when the initial choice happened to have been a goat, which happens 2 out of 3 times, because 2 out of 3 doors had goats behind them. Thus, you have a 2 in 3 chance of winning the Tiffany jewel if you switch; whereas your chances of winning it are only 1 in 3 if you stay with your original door (because 1 in 3 was the probability of picking the correct door in the first place).

If that was hard to wrap your head around, let me raise the stakes some more.

Riddle: Instead of 3 doors, you are faced with 1,000 doors. Only 1 of them leads to the diamond necklace, and 999 of them lead to goats. You have no idea which door opens to the necklace, so you pick a door at random. Monty then proceeds to open 998 other doors to reveal goats, leaving behind only 2 doors shut: the one you randomly chose and the one left closed by Monty. Which door is more likely to lead to the diamond?

Solution: We know that only 1 of these 2 doors hides the massive jewel. One of the doors was chosen by you and the other door was chosen by Monty. The difference is that you have no idea where the diamond necklace is, but Monty does. You chose your door randomly, so it only has a 1 in 1,000 chance of being the correct door; meaning, the other door has a 999 in 1,000 chance of being right! Remember, Monty knows where the necklace is because Monty will never open

> a door that has the necklace behind it—he only opens the goat doors.
>
> If by some chance you chose the correct door from the beginning, then he has free range to open any doors and never accidentally reveal the necklace. However, if you chose a goat's door from the beginning, Monty has to be careful to not accidentally reveal the door hiding the jewel. He has to open every door *but* the necklace door.
>
> Think of it this way: there are 999 bad doors to choose from, and Monty eliminates 998 of them. The door he chooses to leave shut is much more likely to be the door that hides the yellow diamond.

Maybe you're still unconvinced. That's understandable. After Marilyn vos Savant wrote about this problem, she received thousands of letters from readers telling her she was wrong, many of whom were men with PhDs. The problem continues to confuse students to this day.

When in doubt, we can use computers to simulate probability problems like coin flips or Monty Hall games over and over again and settle these debates by seeing the probabilities for ourselves. They may make for fun mind exercises, but in real life we encounter risky decisions that can't be duplicated by a computer.

Subjective Probability

Suppose you hear that a polling organization predicts that Cher has a 75% chance of winning the next US presidential election, or your drag-obsessed friend tells you they think Lil Nas X has a 60% chance of winning *RuPaul's Drag Race* this season. Neither of these events can ever be repeated, much less by a computer, but are we meant to believe that Cher would win

75% of the elections if we ran them infinitely many times? Would Lil Nas X become America's Next Drag Superstar 60% of the time?

Even if it *were* possible to turn back time and repeat the same days over and over, we still couldn't expect to see different, randomized results each time because we also would have needed to duplicate the exact same crop of voters, or RuPaul's exact same frame of mind while judging the competition. Are these events not deterministic?

I noted before that calculating the probability of a queen winning *Drag Race* is as simple as dividing 1 by 12 if there are 12 competitors. But this straightforward math isn't the only way to interpret probability. I'm going to let you in on a little secret: mathematicians don't have a consensus on what probability even is. One leading theory is that it's the relative frequency with which things happen. That definition works for randomized, repeatable events like coin tosses, lottery draws, Pokémon battles, or Monty Hall games. But when there are factors that math can't control, we can supplement probability with educated guesses. This second interpretation is known as a *subjective probability*, which is a measure of belief or opinion. For the pollster it's an opinion based on the reports of a small group of voters, and for the *Drag Race* fan it's an opinion based on preference, and maybe a little experience. When the opinion comes from an expert who has experience and evidence, you may want to believe them, and place your bet on Lil Nas X.

You can see why some people argue every probability is 50%, especially when it comes to nonrepeatable events. But what about when it comes to medicine? Wouldn't you want your doctor to tell you if they thought that a particular drug had a 0.1% chance of causing a harmful side effect and another had a 75% chance? Are these probabilities that state facts about the world, or opinions? On one hand, these don't sound like repeatable, duplicatable events, but on the other hand, we *can* study the effects of a drug on a large sample of people (or rats) and analyze *that* data in lieu of our repeatable event.

Let me give an example. Suppose you develop a new drug that helps relieve headaches, and you name it dragzedamol. To test its effectiveness, you conduct a study on about 5,000 participants who frequently suffer from tension-type headaches and give them your pill, and 59% of participants report feeling less pain after two hours had passed. However, this isn't enough evidence to claim that dragzedamol cures 59% of headaches. You have to account for the possibility of those headaches getting better on their own, so you have to do a separate test on a control group. The control group is a separate batch of 5,000 new participants, and instead of giving them dragzedamol, you give them a placebo, a fake pill that looks identical to yours but has no active medical ingredients. Out of this group, 49% reported feeling less pain two hours after taking the placebo.

To summarize, 59% of people felt better after taking your pill, but 49% felt better anyway; meaning, your pill only made a difference on 10% of the participants. In conclusion, if a random person picked up dragzedamol from a pharmacy shelf, it would only have a 10% chance of curing their headache. This was a real study done by Cochrane, a British charitable organization that organizes medical research. The name of the drug wasn't dragzedamol, though; it was paracetamol, also known as acetaminophen or Tylenol.

In another example from the United Kingdom, a tragic case study of misunderstood probability theory played out in a courtroom. British solicitor Sally Clark lost two babies only two years apart after each suddenly died in their sleep. Sudden infant deaths are a known phenomenon due to natural causes, but to happen twice in the same family within a span of two years aroused suspicion. A month after the death of her second child, Clark was arrested and put on trial for murdering both her children. Respected pediatrician Roy Meadow testified in the trial, claiming that the probability of a single instance of SIDS (sudden infant death syndrome) was 1 in 8,543 for an affluent, nonsmoking family like Clark's, but for two cases of SIDS to happen in a row, he put the likelihood at around 1 in 73 million.

It appeared as though the mathematical likelihood of two natural deaths in a row was all but impossible, and thus the jury voted to put Sally Clark in prison for life. After serving three years behind bars, microbiological evidence found in her son's lung tissue pointed to a death from natural causes, leading to her release. So an innocent, grieving mother, who lost her two infant sons through no fault of her own, was wrongfully imprisoned for years and labeled a child killer by the community, all because of misused probability?

A second look at the case led many to call out all the flawed math done by the prosecution. Let's look at these egregious errors.

Let's assume that Dr. Meadow was correct that the likelihood of a sudden infant death is 1 in 8,543. His big leap from there was to say that the likelihood of two sudden infant deaths happening in a row was equivalent to:

$$\frac{1}{8{,}543} \times \frac{1}{8{,}543} = \frac{1}{72{,}982{,}849}$$

Or about 1 in 73 million. This was incorrect. If you want to know the probability of two events happening, you are only allowed to multiply their individual probabilities if they are *independent*. Independence means that neither event influences the other in any way. For instance, coin tosses, roulette spins, and dice rolls are all independent events, which is why a coin still has a 50% chance of landing on heads even after observing a streak of 99 heads in a row—the past doesn't influence the future. However, when events are not independent, streaks and patterns are more likely to appear. For instance, if you know that it's snowing in Toronto, then it's more likely to also be snowing in Mississauga, the next city over. The weather in one city is *not* independent from the weather in the one near it. Your eye color is independent of my eye color; knowing the color of my eyes will give no useful hint as to the color of yours, unless you and I are closely related by genetics. If you were my mother, our eye colors wouldn't be independent, since your eye color is more likely to match mine.

Dr. Meadow assumed that the deaths of Sally's two children were independent, but the Royal Statistical Society later argued that there could have been genetic or environmental factors that predispose certain families to SIDS, making a second case much more likely. That was Dr. Meadow's first mistake.

Second, even if the 1 in 73 million figure was accurate (which it isn't), it would be wrong to conclude that Sally Clark had a 1 in 73 million chance of telling the truth, as many media outlets put it. This was their line of reasoning:

1. Clark claimed that both her children passed away from SIDS.
2. The probability of that happening is 1 in 73 million.
3. The probability of her innocence is 1 in 73 million.

This is known as the *prosecutor's fallacy.* Let me explain with an example. Suppose that you're walking through the streets of New York City when someone runs up to you and tells you their purse was snatched by a man with red hair, and later you see someone with red hair walk past you. What's the probability that they are the purse thief? Only about 2% of the population has red hair, but this doesn't mean that they have a 2% chance of being innocent. After all, 2% of New York City is over 150,000 people! You could fill up Madison Square Garden many times over with redheads. The probability that the one redhead standing before you is the one you're looking for is actually very low. Sure, it might raise suspicion, but you have to find more evidence than that before coming to a conclusion. Sally Clark wasn't given that courtesy. The system failed her.

Uncertainty can be like a curse. Without perfect knowledge and airtight evidence, we can only make educated assumptions when deciding someone's innocence or guilt. How many mistakes have we made by betting on the wrong things? How much potential tragedy could we avoid if we had better knowledge about the future?

Wicked as it may be, there's no getting rid of uncertainty. We can, however, use math and science to get a little bit closer to accurately understanding events. It may not be infallible, but we can make strong guesses about when a hurricane is near, whether the addition of a new traffic light will cause fewer car accidents, and whether a person is at risk of developing lung cancer. We can use these guesses to decide what games to bet on and when to sit out. Large companies do this all the time. There is an entire branch of math, called actuarial science, that specializes in using probability to manage risks and prepare for the future. Most actuaries work for insurance firms, where their job involves taking huge bets on events like health complications, car accidents, floods, fire, or even death. In order to protect their investments, the insurance company takes a large number of bets on thousands of different customers and relies on the law of large numbers to average out their bottom line. Just as the wins of some Las Vegas gamblers are balanced out by the losses of the majority, an insurance company pays out the benefits of one person using the monthly premiums of many others. Insurance companies never take bets where the odds aren't in their favor. That's why a teenage driver is asked to pay more for car insurance than a driver in their 40s. Insurers seek higher rewards for higher-risk games.

We can take a lesson from their profitable model. As everyday people, we don't always have a say in what games we partake in. Car insurance, for instance, is a legal requirement for car owners in many places. But when we do have a choice, we should only play the games where the odds are in our favor. When engaging in a high-risk game, the rewards better be worth it.

Many people are aware that smoking cigarettes increases your risk of developing cancer, but fewer know that alcohol is a carcinogen, too. Most people who drink on a weekly basis do so as if their only risk is a hangover the next morning. I'm not telling you how to live your life, just that you should be aware of the risks and decide for yourself if the rewards offer enough compensation.

If anyone approaches you with a get-rich-quick scheme that sounds too good to be true, ask yourself, what's the risk? If there is none, you should be even more suspicious. The internet is full of scams that are cleverly disguised as business opportunities, but in reality they are like rigged street games. In big cities like Paris there are organized tourist scams, where a crowd gathers around a man shuffling a ball between three cups, challenging people to guess where the ball is in a bid to double their money. You watch someone bet 50 euros, guess correctly to quickly win 100 euros, and get the impression that the game is easy. There are only three cups, so the chances of winning are 1 in 3; but if you keep your eye on the ball, you figure you must have better odds than that. But what you don't know is that the person who won is actually in on the scheme, and so are multiple others in the crowd. The man behind the cups is using sleight of hand to slip the ball into his pocket so that you can't win. The game looks like it's low risk, high reward, but you're being deceived. Alcohol, get-rich-quick schemes, and street games all have this in common: the risks are greater than they appear.

But I don't want to scare you away from trying new things because I'm a big proponent of exploring outside your comfort zone. There are so many games where the rewards are greater than they seem. When you start a lucrative new job, there's the immediate reward of a higher salary, but you might also make a lifelong friend. When you opt to eat a dish you've never tried before, you might be about to find your all-time favorite food. When you travel to an unfamiliar country, you might enjoy a newfound sense of independence and perhaps even discover the site of your future home. When you lose *Drag Race*, you might incidentally kick off an unexpected and thrilling career. Learning new languages, trying new hobbies, and socializing with new people are all ways to expand your possible outcomes. I think randomness is the spice of life. Tomorrow, you might find $20 on the ground or you might spill a drink all over your beautiful shirt. Who knows what possibilities await you 5, 10, or 20 years from now? You could be a parent, a vegetarian, a saxophone virtuoso, or a drag superstar!

Your life might completely change, if you're willing to roll the dice.

CHAPTER 5

The Average Queen

Who am I? Who am I supposed to be? What do other people see when they look at me? These questions are on our minds from early childhood, when we first begin to measure ourselves against everybody else. When I first moved to Canada, I was 5 years old and the only Filipino person in my class. The rice my mom packed for my lunch was considered so odd by other kids that I'd go home begging for sandwiches instead. I was asked why my skin was brown, why my nose was flat, and why I only played with girls. As I grew into adolescence, I felt so different from other boys that I'd immediately get anxious whenever a teacher would split us up by gender for everyday activities.

Once my classmates got to dating age, I started feeling really left out. People were calling me "gay," "girly," and all sorts of names I won't repeat here. I began to develop negative feelings toward gay boys, especially the effeminate ones. Seeing them on the internet wearing makeup and women's clothing would get on my nerves. As an adult I understand the reason it bothered me so much: my own internalized homophobia. All the negative messages I was being bombarded with about gay people had manifested into feelings of self-hatred and denial that I projected onto others. The truth is that I saw myself in those boys who freely expressed their femininity, and I was jealous of their confidence in breaking away from a status quo to which I was trying my best to conform.

But my experiences have since taught me that, in fact, this "average person" we are all comparing ourselves to doesn't even exist. You may embody the average in some instances, but no matter who you are, there's also sure to be something extraordinary about you.

From Tips to Trees

Although nonexistent in reality, the *idea* of an average person still has a lot to teach us. Averages are crucial to the ways we structure our lives and make decisions because they shape our expectations and limits. For instance, comparing the average worker's income to the average price of rent for an apartment in a particular area can provide useful insights about housing affordability, and comparing the average salaries of different demographics in the same profession might alert us to the possibility of systemic inequality. Averages can be used to spot patterns, trends, and problems in the broader world, and in our daily lives.

For instance, someone who earns tips may be cued to investigate if they notice days when their tips don't seem in line with what they usually expect.

This is what I made in tips from my last 6 drag shows:

$180, $125, $65, $210, $195, $200

The average among these is $162.50. This type of average is known as the *mean*, which is calculated by adding together all the numbers and then dividing by how many are in the set. If I needed to predict how much money I might make from a future show, I could fairly use $162.50 as my benchmark, even though I've never made exactly $162.50 on any night. Actually, my tips for one evening almost always total a whole number, since the smallest paper denomination in Canada is a $5 bill, and I don't appreciate people throwing coins at me!

The "average night" in tips that I calculated never existed in actuality, just like the "average human" isn't out there roaming the streets. Rather, an average is simply a hypothetical idea distilled from a larger pool of data. It is worth mentioning that the average *can* be a part of the data set. For example, the mean of the three numbers {10, 20, 30} is 20, which you can see is also a number in the set. But the average isn't *necessarily* part of the data set. Every drag performer counting their tips at the end of the night is thinking like a statistician. If the tips are significantly greater than average on a particular night, it may signal that your new Whitney number was a crowd-pleaser, and you should work "I Wanna Dance with Somebody" into your regular rotation!

There's more than one type of average; another is known as the *median*. The median of a group of numbers arranged in numerical order is the middle number. For example, the median of the set {180, 125, 65, 210, 195, 200} would be 187.5. We calculate it by reordering the set from least to greatest: {65, 125, 180, 195, 200, 210}, and if there's an odd number of data points, we would simply select the middle number. But since there isn't one in this case, we take the mean between 180 and 195, which is 187.5.

Another kind of average is the *trimmed mean*, where you throw away the top 10% and bottom 10% of the data and then calculate the mean of what's left over, in order to discard any extreme outliers.

The decision of which type of "average" to use can be a subjective one.

Suppose for instance that two music fans are in an argument over their respective favorite artists: PSY and BLACKPINK. They decide to settle their argument by comparing the average number of views for each artist's top 5 music videos on YouTube. As of 2022, their top 5 videos raked in these view counts respectively:

PSY: 4.6 billion, 1.5 billion, 819 million,
664 million, and 383 million.
BLACKPINK: 1.9 billion, 1.7 billion, 1.5 billion,
1.3 billion, and 1.2 billion.

The PSY fan might insist, "PSY is obviously more popular because none of BLACKPINK's videos have even close to 4.6 billion views! On average, PSY's videos have 1.59 billion views, while BLACKPINK's videos only have an average of 1.52 billion views." However, PSY's music videos only have more views based on the *mean*, not on the *median*. If we go by median views instead, BLACKPINK wins, with an average of 1.5 billion views compared to PSY's measly 819 million.

PSY's videos have a higher *mean* view count due to the inclusion of a break-out song you might have heard once or twice called "Gangnam Style." Means are often tilted toward large outliers, and "Gangnam Style" was a viral hit—the first video on YouTube to ever hit 1 billion views! The record-breaking 4.6 billion views for that video pulls PSY's mean higher, but his other music videos haven't had the same success. BLACKPINK's top video "Ddu-Du Ddu-Du" may not have gone as viral as "Gangnam Style," but their hit-making prowess is more consistent.

We often use medians when describing average incomes. A small group of mega billionaires would raise the *mean* income of a large group of earners, but not the *median*. In Canada, the median household after-tax income was $66,800 (Canadian dollars) in 2020, meaning that half of households took home more than that and half took home less. The median will always be the "middle number" in any group of data, but the mean is only the middle number when the data is perfectly symmetrical. Sometimes it's not completely obvious which metric to use, and in this sense statistics is like an art. The decision is up to the statistician and how they want to present the data—are they a PSY fan or a BLACKPINK fan?

In our daily lives we conduct mental analyses to find averages all the time, even about things that have nothing to do with numbers. Try to picture what the average drag queen looks like. Do you envision a tall goddess wearing false lashes, high heels, and sequins? Are they wearing blond hair or brown? Are they tan, dark-skinned, or fair-skinned? If we were to use the mean, putting all drag queens into a blender and mixing them up, your average queen would most likely be a 200-pound pile of flesh, synthetic hair, and microplastics. Or at least this method is how the YouTuber Safiya Nygaard found the color of the average Sephora lipstick. She bought one tube of every available color (603 to be exact), melted them all together in a pot over heat, mixing in all the nudes, pinks, reds, browns, purples, and the few bold neon blues, greens, and oranges. When she poured the blended mixture back into lipstick tubes to cool off and solidify, she found the average Sephora lipstick color looked like a dark mauve-y purple, with a subtle satin finish.

Luckily, in real life we don't have to go to all that work. An average is usually just something you feel by intuition. Being able to form a mental average using past experience is what makes adults different from children. It's the wisdom of experience that gives us the power of insight. Evidence of this intuition can be found in one of the oldest, longest poems ever written, the ancient Hindu epic, the *Mahabharata*, which may have been created as long ago as 400 BCE. In one of the major plot lines, King Nala loses his domain over a gamble with his brother. Hoping to win his kingdom back, he asks for help from King Rtuparna, a master at dice. Rtuparna takes Nala to a large vibhitaka tree and informs him that two large branches of the tree contain 2,095 pieces of fruit, inviting Nala to count them one by one. Rtuparna made the estimation by first looking at the fruits that fell on the ground and then extrapolated to the two large branches and, potentially, the entire tree. Rtuparna is showing off a skill known as *sankhya*, a Sanskrit word that means "to count, reckon, calculate, or reason," an important skill among farmers in India who often needed to roughly appraise their crops. According to the epic, making large estimations using small measurements has useful connections to gambling and probability. At first glance, a game of dice appears to be based on luck alone, but armed with a new mastery of counting, Nala is able to turn the odds in his favor and win his kingdom back after challenging his brother to a rematch.

Sometimes there are just too many pieces of fruit on a tree for us to invest the time to count them one by one. But often we want information on *all* pieces of fruit, or on all Canadians, all drag queens, or all Sephora lipstick colors. When counting every single one is too much work, counting a small sample can be enough to extrapolate an average, just as Rtuparna did. We have randomness to thank for this shortcut.

Card Games of Mass Destruction

In the previous chapter, we looked at randomness and saw how we can use math to deal with the chaos, buffoonery, and uncertainty of life. But randomness can be leveraged to solve important problems. Let's look at a classic example.

The game of Solitaire (also known as Klondike, Patience, or Canfield) is a single-player game you might indulge in on your computer when you're bored and trying to avoid work. The game begins with a shuffled deck of 52 playing cards, and the goal is to sort them into order by suit and rank. However, you can only move cards which have been flipped face up. Since most cards are face down at the beginning of the game, the ways you can move the cards' position are very limited. But each time you move a card, it flips another card face up, which can thus lead to more possible moves. If you run out of possible moves before completely sorting the deck, you lose.

Despite being a simple game, losses are common; even a computer that's programmed to play perfectly will lose some games. Sometimes it's the deck's fault. If the cards have been shuffled in an unfavorable order, a game can be impossible to win. And that leads us to an interesting question.

> ***Riddle:*** How many unwinnable Solitaire shuffles are there? What is the probability that any random shuffle of the deck will make for an unwinnable game?

> ***Solution:*** So far, there is no solution. This is an unsolved problem in mathematics, which is hard to believe because it feels like mathematicians have solved far harder problems than this one!

In theory, we should be able to figure this out. All we have to do is count the number of unfavorable shuffles, and then count the number of ways to shuffle a deck. But checking every possible shuffle is just too much work due to the sheer number of possible shuffles that exist. Remember in the previous chapter, when we calculated the number of ways we could shuffle six lottery numbers? The math for shuffling cards is the same: there are 52 possible cards that can appear on the top of the shuffle, and then there are 51 remaining cards for the second position, 50 for the third, and so on. The formula we're talking about here is a *factorial*. There are 52! ways to shuffle a deck of 52 cards.

$$52! = 52 \times 51 \times 50 \times 49 \times 48 \times 47 \times 46 \times 45 \times \ldots \times 5 \times 4 \times 3 \times 2 \times 1$$

That works out to a total of 8.06×10^{67}, or

80,658,175,170,943,878,571,660,636,856,403,766,975,289,505,440,883,277,824,000,000,000,000

This comes out to just over 80,000 vigintillion possible card shuffles, or 8×10^{67}, which is mathematical shorthand for writing "8 followed by 67 zeros."

How big is this number? Suppose you asked a computer to check all 80,000 vigintillion shuffles, and play all 80,000 vigintillion possible games of Solitaire, keeping a counter of how many of them lead to unwinnable games. Even if your computer were capable of checking 100 billion shuffles per second, it would still take 2,555,903,337,736,199,158,733,890, 944,064,306,758,919,864,167,138 years to check every shuffle. So you have a lot of time to burn while the computer does the work.

While waiting for the millennia to pass, you and your immortal monkey sidekick might decide to play the lottery. (With infinite cash, squandering it on games of chance isn't such a problem.) Ontario's Lotto 6/49 wouldn't be a bad way to go; as you may recall, it gives you about a 1 in 14 million probability of winning. You buy one lottery ticket every billion years and, figuring you might as well get some exercise while you're at it, every time you win, you step forward by one centimeter. Continue this until you have walked the entire circumference of the earth. Once you're done circling the globe, remove one grain of sand from the Sahara. Then repeat this all over again and remove another grain of sand each time you arrive back where you started. When the Sahara is empty, move on to the Arabian Desert. All this time, your computer is chugging away speedily, checking 100 billion shuffles per second. Do this over and over again until you've cleared out all the grains of sand in all the world's beaches and deserts, at which time the progress bar on your computer will tell you that it's about 0.2% of the way there.

Checking every possible shuffle of the deck is what computer scientists would call an *intractable problem*. Sure, it's possible to solve, but you won't be getting an answer any time soon. Theoretical math hasn't found a solution to this conundrum, either, but maybe you can be the mathematician who changes that!

Here's where randomness comes in. The mathematician Stanislaw Ulam came up with this great idea: just play a lot of Solitaire. Maybe not vigintillions of games, but play hundreds of games, thousands even, all starting with random shuffles, counting how many of these games are unwinnable. Ulam's experiment landed on an answer of around 18%: approximately 18% of all shuffles lead to unwinnable Solitaire games. Instead of counting all the fruit on the tree, Ulam looked at only a single branch. In this case, the branch was a small subset of possible shuffles; the shuffles he used in his experiment were only a tiny fraction of all 80,000 vigintillion shuffles that are possible. Whenever we're dealing with a mountain of data, inspecting every individual element is often wildly beyond our capabilities, so it's a lot more practical to just use smaller, random samples. We can leave it up to chance and still get a pretty good estimate.

Statisticians make an important distinction between a theoretical, true value and the value they distill from experimentation. The true probability that a random shuffle of a deck will produce an unwinnable game of Solitaire will never be known exactly until that computer finishes calculating, some quindecillion years from now, but we can derive an experimental probability, which is around 18%. So how close is 18% to the exact answer?

Imagine that a mysterious drag queen randomly emerges from a cave in front of you, hands you an odd-looking coin and asks you to find the probability that the coin will land heads up. You toss it 4 times, and it lands heads up twice (that's a 50%

heads rate). You toss it 100 times, and it lands heads up 41 times (that's a 41% heads rate). You toss it 1,000 times, and it lands heads up 438 times (that's a 43.8% heads rate). You may be convinced by now that this is a weighted coin that slightly favors tails, but you're unsure about the exact probability.

What answer do you give the drag queen? This problem is identical to Ulam's Solitaire problem, where the true probability of landing heads up or losing Solitaire is unknown, floating around somewhere in the ether. Perhaps you'll say that the probability is somewhere around 44%. This is known as a *point estimate*, a single value that is your best guess of an unknown value. Or, you might respond by saying that the true probability is somewhere between 40% and 50%. This is called an *interval estimate*, a range of multiple values.

Point estimates are good heuristics (meaning mental shortcuts), but interval estimates can be much more powerful. When statisticians repeated their Solitaire experiment multiple times over, they found that 95% of the time, the proportion of unwinnable shuffles was between 17.948% and 18.140%; so they can say with 95% confidence that the true probability lies somewhere in that interval.

Statistics gets even harder when we leave behind hypothetical scenarios like coin tosses or card shuffles. Humans are much more unpredictable, and our biases get in the way of conducting proper statistical studies fit with truly random samples. For example, many scientific studies throughout history have been conducted using only male subjects—from testing car crash dummies that mimic average male bodies to testing male rats in biomedical research—because researchers presumed that male test subjects were representative of the entire population. This has resulted in, for example, women receiving improper medical care when their symptoms differ from the typical ones exhibited by men. In the 1990s, Bernadine Healy, a cardiologist and the first female director of the

US government's National Institutes of Health, coined the term "Yentl syndrome" to describe the underdiagnosis of heart disease in women.[1] This phenomenon is named after the protagonist in Isaac Bashevis Singer's short story about a girl who dresses as a boy in order to receive an education. (It was the inspiration for the stellar 1983 film starring the incomparable Barbra Streisand in drag.) Dr. Healy writes that women must present the same physical symptoms as men in order to be taken seriously by medical professionals because the majority of statistical studies have only collected data on men. The average *person's* warning signs and symptoms of heart attack turn out only to be the average cisgender *man's*.[2]

Data collection is serious business, and there are additional hurdles when collecting data on queer people due to the social stigma attached to "coming out." For instance, a fair question that a policymaker may be interested in asking when attempting to assess legal protections for queer people is, "What percentage of Americans identify as LGBT?" In 2022, Gallup conducted telephone interviews with 10,736 people in the United States and found that 7.2% of respondents identified as lesbian, gay, bisexual, transgender, or some other color in the LGBT rainbow.[3]

I find it fascinating how different the responses were when sorting by age brackets. Among the oldest demographic, those born before 1946, only 1.7% of respondents identified as LGBT, compared to 19.7% of Gen Z respondents (those born between 1997 and 2004). There's sure to be some margin of error when conducting telephone interviews; some participants may not feel comfortable coming out to a stranger over the phone, and others may be confused by the wording of the questions. The 7.2% figure may not accurately capture the *true* percentage of LGBT Americans, just as mathematicians haven't accurately captured the true probability of losing Solitaire, but the important thing is that random sampling is close enough for us to draw practical conclusions. One conclusion that Gallup researchers came to was that, barring a reversal of the increasing comfort levels of LGBT people to identify as their authentic selves, the LGBT proportion of the US population can only be expected to grow in the future.

Statistical Influencers

Statistical studies can influence laws, hiring quotas, school curriculums, and access to medicine. One of the greatest pioneers in advancing statistics in medicine was a British nurse named Florence Nightingale. She was born in 1820 to an affluent family who strongly endorsed her education but never wanted her to become a nurse. At the time, nursing was seen as beneath a lady of her social status. (It could have been worse—she could have wanted to become a drag queen.)

As strong-willed young people often do, she rebelled against her parents, enrolling in nursing school despite their objections. In 1854, she and a team of volunteer nurses were sent to Constantinople (modern-day Istanbul) to aid wounded soldiers in the Crimean War. What they found was a hospital without beds, blankets, or furniture, with patients lying in their own waste, amid swarms of rats and fleas. At the time, it was not yet fully understood how germs could spread disease, and so standards for hygiene and sanitation were far lower than they are today.

Nightingale recognized the connection between germs and disease, and her faith in her conviction changed the world.

She and her team started by cleaning the entire hospital, requiring proper hygiene from both the nurses and the patients, from clean linens to hand washing. She and her team also kept meticulous records on the soldiers they treated and their causes of death.

Upon her return to Britain, she presented her findings drawn from these data, along with recommended medical reforms. She constructed what's known as a rose diagram, which looks like a pie chart with slices of different lengths showing causes of death among her patients over her two years in Crimea.

The diagram revealed two main takeaways: (1) more soldiers were dying from preventable diseases, like cholera or typhoid (the gray wedges in the diagrams), than from battle injuries (the light-colored wedges toward the center); and (2) the

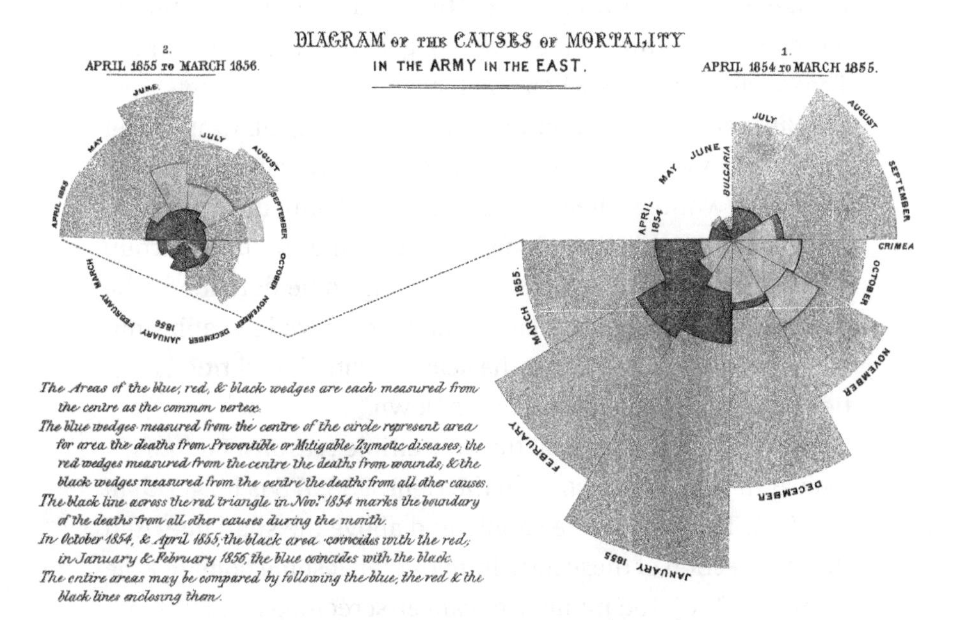

Causes of death during the Crimean War. The right diagram shows the year 1854–1855, and the left diagram shows the following year, 1855–1856. The lightest color (toward the center) represents deaths from battle injuries, gray (the largest area) represents deaths from disease, and black represents deaths from all other causes.

Source: Florence Nightingale (1820–1910), "Diagram of the Causes of Mortality in the Army in the East," published in Notes on Matters Affecting the Health, Efficiency, and Hospital Administration of the British Army, *c. 1858.*

deaths from preventable diseases drastically decreased after Nightingale's team implemented new hygiene and sanitation standards (the diagram on the left shows the gray wedges decreasing in size over time, represented by moving clockwise around the circle). Remember, the British were only beginning to learn about how germs spread diseases. Nightingale didn't convince officials using biological studies—she used statistics. Her reforms were responsible for saving countless lives, and her name lives on; she is often called the founder of modern nursing, and newly graduated nurses take the Nightingale Pledge just as doctors take the Hippocratic Oath. Inspired by her courage, young women soon began aspiring to be nurses. That's what happens when you have diverse role models!

But beware: not all graphs are created equal. Graphs can be manipulated to suit the purposes of the creator. In 2015, the American women's health organization Planned Parenthood was put on trial before Congress after an anti-choice group made a string of videos calling for them to be defunded. The Republican congressman Jason Chaffetz showed a graph similar to the one on page 121 at the hearing, with the subtitle "Abortions Up—Life-Saving Procedures Down."

This graph is problematic for multiple reasons. First of all, look at the end of the lines on the right: 935,573 is bigger than 327,000. The dotted line should end above the solid line. And the intersection of these lines implies that at some point in time (around 2010), the number of cancer screenings and prevention services exactly equaled the number of abortions at Planned Parenthood. But a closer look at the numbers on the graph shows that, if depicted accurately, the lines would never touch at all. The slope of the lines also tells a story about how fast quantities are changing, and the jump from 289,750 abortions in 2006 to 327,000 in 2013 is a much smaller increase than the decline in cancer screenings and disease prevention services (2,007,371 down to 935,573 is a decrease of more

Planned Parenthood Federation of America:
Abortions Up–Life-Saving Procedures Down

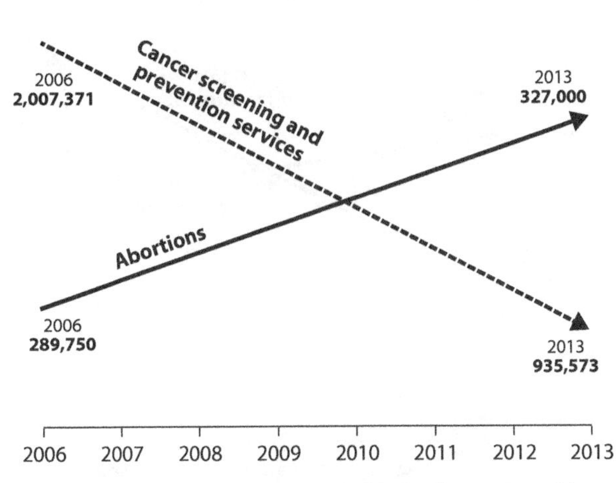

A chart shown by US Representative Jason Chaffetz, a Republican from Utah, at a congressional hearing on September 29, 2015, investigating Planned Parenthood.

Source: Linda Qiu, "Chart Shown at Planned Parenthood Hearing Is Misleading and 'Ethically Wrong,'" PolitiFact, *October 1, 2015, https://www.politifact.com/factchecks/2015/oct/01/jason-chaffetz/chart-shown-planned-parenthood-hearing-misleading-/.*

than 1 million). So the line representing abortions should look much flatter in comparison. This is all given a veneer of legitimacy because the graph doesn't have a vertical axis, so the lines are drawn with no context, other than to show that up means rising and down means falling.

You might say that the overall message of the graph remains the same: the number of abortion procedures is rising while life-saving cancer screenings are falling. Even leaving aside the fact that abortion certainly can be a life-saving procedure, this analysis is guilty of yet more bias by omission because it ignores the other work done at Planned Parenthood. On the same day, *Vox* created and published a revised graph (see page 122) which includes a labeled vertical axis, data from every year between 2006 and 2013, and data on other services provided by Planned

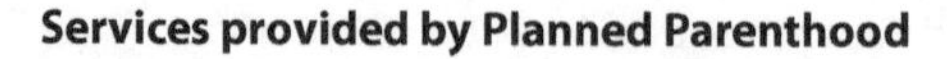

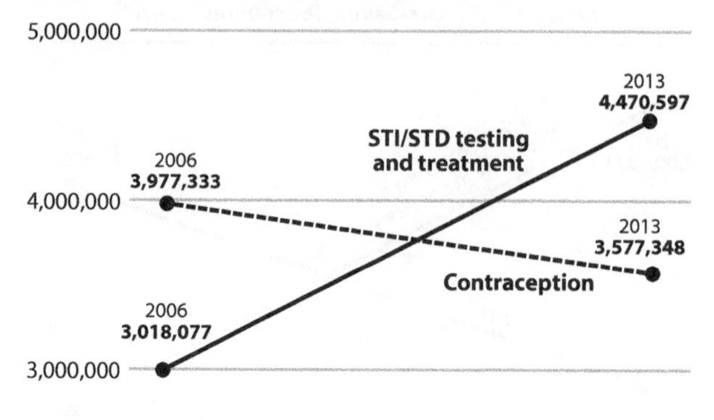

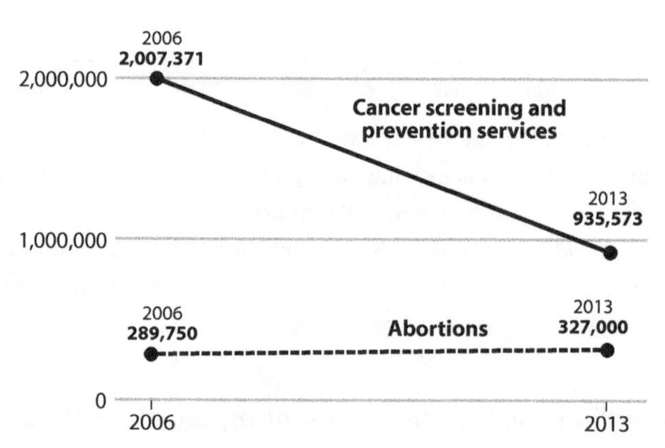

A chart made by *Vox* that more fairly depicts the range of services at Planned Parenthood.

Source: Timothy B. Lee, "Whatever You Think of Planned Parenthood, This Is a Terrible and Dishonest Chart," Vox, *September 29, 2015, https://www.vox.com/2015/9/29/9417845/planned-parenthood-terrible-chart.*

Parenthood, including providing contraception and testing for and treating sexually transmitted infections.

Clearly, this graph is not nearly as provocative and, crucially, it shows the big picture, not the cherry-picked statistics Chaffetz wanted to leverage.

Statistics can be an incredibly powerful tool, but when statisticians become overzealous or myopic, their craft can easily turn dangerous. One of the founding fathers of modern statistics, Francis Galton, is an undeniable example. Galton was born into a wealthy English family with a strong interest in science. His grandfather was Erasmus Darwin, the illustrious doctor, polymath, and fellow of the Royal Society. Yes, *those* Darwins. Charles Darwin was Galton's half cousin.

Galton began his scientific career with an expedition to Africa, in what is now Namibia—an area of the world almost entirely unknown to Europeans at the time. His extensive writing on the Tribal peoples he encountered there and his pioneering cartographic work in the region earned him a gold medal from the Royal Society on his return to Britain.

Around the same time, his cousin Charles was in the midst of writing *On the Origin of Species*, which would become the foundational text in evolutionary biology. Not only did it change the way science viewed the natural world, but it also altered the course of Galton's life. Darwin's work on natural selection demonstrated that we can breed plants and animals to have desirable traits through artificial selection. Galton wondered if such manipulation might be possible for humans through selective breeding—allowing only the most "desirable" humans to have offspring, while ensuring that those perceived to be "faulty" remained infertile.

His studies required new statistical techniques to contend with his large data sets, so he invented them. Some of the techniques Galton developed are at the bedrock of the field and are used in every statistics course through the present day. Among these is the concept of *standard deviation*, which measures a data point's average distance from the mean. Just as I can calculate the mean value of how much money I might make in tips in one night, I can also compute the standard deviation. If my data points (in this case, the amount of my nightly tips) are similar, they will be clustered together on a graph, indicating a low standard deviation, whereas a high standard deviation is a sign that the tips varied greatly from one night to another.

Galton attempted to use this and other statistical tools he developed to quantify human excellence and discover whether it was hereditary. He called his new field of study *eugenics*, which he defined as "the science which deals with all influences that improve the inborn qualities of a race."[4]

There are obvious issues with asking people to self-report how "excellent" they are, so Galton had to come up with his own definition of human excellence. In one study, he looked at a sample of 286 highly respected judges and found that 109 of them had one or more "eminent relations." Those 109 judges could be grouped into only 85 families. There were more familial relationships among those 286 judges than one would find in a random sample of 286 people in the general population, and Galton saw this as evidence that excellence is inherited.

Galton decided that his next logical step was to associate intelligence with race and, to no one's surprise, his data bore out that he himself was part of a superior race of gifted individuals. On this basis he went on to advance some terrifying ideas. His 1883 book, *Inquiries into Human Faculty and Its Development*, insisted that the extinction of an entire race was not problematic, so long as it was brought about "silently and slowly through the earlier marriage of members of the superior race."

In his letter to the editor of *The Times*, he also proposed that "industrious, order-loving Chinese" be resettled in Africa, which he lamented was "occupied by lazy, palavering savages."[5] Galton was obviously deeply racist and just plain wrong for a number of reasons, one of them being that collecting unbiased data about excellence and giftedness is impossible because of how hard it is to define those qualities. It's one thing to collect data on heights, ages, incomes, and even sexualities, but it's extremely difficult to quantify measures like intelligence. Should intelligence be based on memorization ability, speed and accuracy of mental math, understanding of social cues, knowledge of general trivia, or something else entirely?

Intelligence is far too complicated to be captured by a single number, and it's so subjective that the definition of intelligence is left up to the test maker. Yet IQ tests have become widely popular since the time of Galton, and the notion of having a high or low IQ has made its way into every corner of culture. IQ tests were originally introduced in France to try to identify struggling children who, when given more help in school, could become successful. When the intelligence test came to the United States, it was used to screen army recruits. The test subjects were measured by standards set by white American men, even though many of those being tested were minorities who did not receive the same formal education white people did, or were immigrants who barely spoke English. These tests reaffirmed racist beliefs that some races were on average intellectually superior to others. Biased data gathering leads to biased results.

Policymakers have long used faulty statistics to justify subjugating entire groups of people. Theodore Roosevelt feared America would commit "race suicide" if it didn't stop Black and brown people from supplanting the supposedly superior American race.[6] Sound familiar? During his time in office, he defended the US colonization of the Philippines, believing that Filipinos were savages incapable of self-governance who needed to be taught democracy.

Elsewhere in the country, the American Eugenics Society traveled around to state fairs to encourage affluent Americans to have large families, while also advocating that the "socially inadequate" should be sterilized. When 17-year-old Carrie Buck was raped and impregnated by her foster parents' nephew, her foster parents blamed her, promptly committing her to the Virginia State Colony for Epileptics and Feebleminded.

Eugenicists rallied, using Carrie Buck as the prime example of someone who should be forcibly sterilized. The state of Virginia agreed. When Buck fought her legally appointed fate in front of the Supreme Court, the judges ruled 8 to 1 to uphold Virginia's eugenics-based law. Carrie Buck was the first of more than 60,000 Americans (we know of) who have been legally sterilized against their will, and a disproportionate number of them are Black women.

Up north in Canada, intelligence tests were used to categorize people as mentally deficient, and the province of Alberta passed the Sexual Sterilization Act in 1928 to allow these groups to be sterilized against their will. Indigenous women were among the groups who were disproportionately targeted for sterilization. The act was repealed in 1972, but we know that Indigenous women in Canada were still being sterilized against their will until as recently as 2017.

There's no question that Francis Galton made some huge contributions to math and science, but he also laid the groundwork for some of the most heinous acts of genocide in the modern world. His promise of leveraging statistics to ensure a future filled with only beautiful, intelligent, and able-bodied people attracted many of the world's most famous and well-respected thinkers, scientists, activists, and leaders. It trickled into the homes of average Brits, Americans, Canadians, Australians, and much of Europe, too. It was only after eugenics provided the Nazis with justification for the Holocaust that interest among the general populace began to wane. But it still has its supporters today among white nationalists, neo-Nazis, and white supremacists. Far-right extremists continue to spread fears that the "woke mob" wants to supplant the white race, and they spread their ideas to millions of young followers online. I hope that the world has learned from its mistakes.

Every statistician must be aware of potential sources of inaccuracy in their analyses that can stem from biased samples, poorly worded questionnaires, dubious assumptions, or even mistakes in data entry. Crucially, a statistician must also take ethics into account. One has to carefully consider the kinds of actions their statistical practices could justify. Even if numbers don't discriminate, people do. Statistics, perhaps more so than any other branch of math, requires human oversight, from deciding which average best describes a data set and which sample best describes the population, to which graphs best summarize a statistical study.

These decisions require some degree of subjectivity and are thus susceptible to bias. Proactively incorporating and supporting the voices of marginalized groups is one way to safeguard against prejudice. As powerful as statistics can be for good, they can be twisted for selfish reasons, which underlines the need for skepticism. Whenever you see any statistic, question the source. What is being implied? What was the methodology? What was the context of the study? Who is defined as average, and as a result, who becomes an outlier?

Distilling a large data set into a single number or graphic is an incredibly powerful and useful tool for identifying patterns of inequality or for predicting the future, but at the same time, it washes away a wealth of complexity and diversity. This can perpetuate glass ceilings, or even risk the erasure of all of those who are considered statistical anomalies.

We may be on the brink of entering a new technological age, one where artificial intelligence is automating facial recognition software, self-driving cars, military drones, job-hiring algorithms, and even academic research. It's important to remember that even our future robotic overlords are trained on data that is curated and created by humans, and therefore are just as susceptible to bias as we are. Mathematical tools can be the most powerful influencers of all, and we must use them with care and intention.

CHAPTER 6

Growing Pains

My life can be characterized as a stream of growth, loss, and change. No one's path is a straight line without its fair share of twists, turns, back flips, flip flops, and spread eagles. Tiny decisions can compound into giant consequences. Some changes creep up slowly, then gain speed at a thrilling (or terrifying) rate. My own evolution as a drag queen has raced along at a good clip and has taken me places I never expected.

The first time I went out in drag, I dressed as a certain animated sea witch for a Halloween party, with purple paint covering my body, and crowned by the same ratted, gray wig I had used for my Faustian book report in high school. That wig has seen some things! I wore a black dress from the thrift store, tucked into black leather pants. Girl, I looked ridiculous. But I felt stunning, and that's all that mattered.

Emboldened by my experiment, I kept practicing my makeup skills, and before long I was teaching myself how to style wigs and sew dresses. One year later, to the day, I sashayed onto the stage for my first official drag performance. My university bar hosted an amateur drag night, which I slayed with a lip sync to "Turn Me On" by David Guetta featuring Nicki Minaj, serving a glam-zombie look and once again rocking my black leather pants. (They weren't my signature piece or anything. I just had nothing else to wear and cheap taste in fashion.)

Fast-forward to October 2022, which I spent as the face of a national beauty campaign for Quo Beauty's Halloween makeup collection. It was a pretty big leap from the local pub to having my picture in drugstores all across Canada. I now wear much more expensive costumes and earn much higher rates, but I still have the same cheap taste in fashion! My guiding principle in drag is maximalism: before you leave the house, always take a look in the mirror and then put on one more thing.

Sometimes I think about that ethos in contrast to my dad's example. He was extremely frugal my entire life. He wore the same set of clothes all the time and never bought anything new or trendy. At the grocery store, he would mentally calculate which cuts of meat were the best value, and he and my mom would check the receipts line-by-line to make sure we weren't overcharged for a thing. If there was even one cent out of place, he would turn the car around, march up to the cashier, and politely ask for the penny back. He was teaching me about mortgages and debt by the time I was 10 years old.

He loved to sing and be silly like I did. My signature song was "My Heart Will Go On," and his was "My Way." But as soon as I began feeling different from other boys, I started growing apart from him. Once, we were at the mall with some family friends, and he refused to let me buy a shirt I had picked out because it was "too colorful for a boy." I protested that it wasn't from the girls' section; it was just a blue shirt with a bright logo on the front! He explained that he didn't want me to wear anything that even slightly resembled a Pride flag because the kids at school might bully me for it. But at that moment, I only felt bullied by him. I would have rather had the whole world turn against me than feel like my own parents didn't accept me for who I was.

Let me give some advice to any parents reading this. When kids don't feel understood by their parents, they go searching for someone who will make them feel understood, and they'll do it behind your back. In my case, I found a sympathetic community on a site called Tumblr. I started sharing my story at 12 years old with people much older than me. Honey, I was fully opening up to strangers and telling them secrets that no one else in my life knew. A stranger on Tumblr was the first person I ever came out to as gay. The anonymity of it all was like a safety blanket. There was no consequence if a stranger on the internet found out I was gay. There was no one there I could disappoint.

I was lucky to get safely out of an online world that can be dangerous for young users. Homophobia drives queer people underground, forcing many of us to find new families, create new identities, and learn how to find love all on our own, with no parents to teach us, no generational wealth, and no safety net. I was fortunate that my parents ended up being very supportive of me becoming the gay icon that I am!

After my dad was diagnosed with cancer, he started to become sentimental in ways he never was before. He had always been so serious and reserved with me. When he told me about his diagnosis, it was one of the only times I ever saw him cry. He would pull me aside and tell me, "If anything ever happens, always remember that I love you." I was too young to process that his behavior was out of character because he was dying. He started reading all sorts of books and blogs about the meaning of life, and concluded that the answer to all our problems was minimalism: living with few material possessions so that we could make more space for living in the moment, and focusing on enriching our lives with real connections and experiences instead of consumerist "stuff." This was kind of awkward because I was over here secretly wanting to become a drag queen maximalist.

He thought I was becoming too materialistic, spending all my money on makeup and jewelry. He was also suspicious of online shopping because he didn't trust any website to take his credit card details, which was bad news for me as an aspiring YouTube beauty guru. As his health was declining, one of his biggest concerns was saving aggressively and paying off our mortgage so that our family wouldn't be in financial trouble if he were gone. I was never able to tell him the news that had I won a scholarship covering my entire university tuition. The last thing he said to me was, "Stay true." I remember weeping by his bed as he drew his last breath, and then my mom and I went shopping for the black suit that I would wear to both his funeral and my prom.

After my dad died at only 44 years old, I had to grow up fast and become an independent woman. It wasn't until after he was gone that I realized I had absorbed all the little lessons he tried to teach me, like, "Never carry a balance on your credit card," and "Don't go into debt just to buy things you don't need."

Why was he so obsessed with debt? I didn't really understand until I went to university to study mathematical finance. That was where I learned how debt makes the world go 'round. The whole economy is a big system of borrowing and lending money. People borrow money from banks in order to pay for houses, cars, college tuition, or to start a business. Governments take on debt, too, to finance projects like highways, trains, airports, or to pay for social programs like healthcare or education. Small interest payments can compound over time to turn a small debt into an overwhelming, unmanageable one.

Here's a simple example of the way interest can grow:

> **Riddle:** Suppose a savings account has an interest rate of 10% per year, and the current balance is $1,000. With no additional deposits, what will the balance be in 20 years?
>
> **Solution:** Let's start small and count the first few years, then see if there's a pattern or formula we can

use to quickly get to year 20. First of all, 10% of $1,000 is $100, so in year 1, the balance will grow by $100, to a total of $1,100. In year 2, that will grow by another 10%. A more efficient way of calculating a 10% increase is to add 1 to the percentage: 10% expressed as a decimal is 0.10 (or simply 0.1), so an addition of 1 makes 1.1. To find a 10% increase on $1,100, we just have to multiply $1,100 by 1.1, which is $1,210. That's an increase of $110 from last year. In year 3, we have to multiply by 1.1 again, giving us $\$1{,}210 \times 1.1 = \$1{,}331$. Each additional year, we multiply by another 1.1, and we can summarize this by using exponents. The number of years, *n*, is the same as the number of times we multiply by the growth rate, 1.1:

$$1{,}000 \times (1.1)^n$$

To find out the balance in 20 years, we simply have to replace *n* with 20.

$$1{,}000 \times (1.1)^{20} = \$6{,}727.50$$

So, 10% growth over 20 years is enough to make the balance grow over 6 times in size! What if we wait an additional 20 years? Will the balance double from nearly $7,000 to $14,000? Doing the math here, $1000 \times (1.1)^{40}$ gives us a whopping $45,259.26, which all started from an original $1,000 growing at 10% per year.

Exponential Growth

The reason that money accumulates interest can be explained by a concept called the *time value of money*. Imagine this: if you were given the chance to either win $10,000 today or $10,000 30 years from now, which would you choose?

I'm willing to bet a large majority of us would rather have the money now. We've all got bills to pay and wigs to buy, and who knows what life will be like in 30 years, anyway? The future is uncertain, and that's why money has a time value. In order for the choice to be more attractive, we'd need to be offered a larger sum of money in the future, like $100,000 in 30 years versus $10,000 right now.

The interest is the extra incentive you get for waiting longer. You can also think about it like the cost of borrowing money. If you borrow $1,000 from a friend or family member, in addition to paying them back, you might also treat them to a meal or a little gift. You end up paying back more than what you borrowed.

In real life, it's much easier to accumulate huge debts than it is to accumulate huge profits, as many a *Drag Race* contestant found out to their dismay. The interest rates we pay on debt are vastly higher than the interest we can hope to earn by saving our pennies and investing carefully. Debt is like a fire that can warm you for the night but grows quickly out of control if you're not careful.

In the United States, there are 44.7 million Americans with student loan debt averaging $32,000 per person, and it's taking many of them decades to pay it off. And it's not just students struggling to satisfy debts, either. The average American family owes over $6,000 on credit cards. For many families, no amount of budgeting or investing will be enough to lift them from a debt-driven cycle of poverty. If it seems like the system is rigged against us, it's because it is.

To make matters worse, stereotypes exist in many parts of the world that assure us finances are best left up to men—straight men in particular. My friends joke that I took a "straight man's major" by studying finance. To me, that's worse than calling me a slur!

If we leave all the financial decisions up to straight men, where does that leave the rest of us? In the United States, women weren't even allowed to open their own checking accounts until 1960, and it wasn't until 1961 that the Indian government prohibited the use of dowries as a monetary transaction from the bride's family to the groom's upon marriage. Despite the law, some families still expect dowries to this day, putting women under a significant financial burden and sometimes resulting in dowry-related crimes such as fraud, theft, and even murder.

In 1976, Ireland finally passed an act that acknowledged married women as co-owners of their family homes. Prior to that, a husband could sell the family home without having to consult his wife. Unfortunately, this law didn't extend to same-sex couples, who didn't even have the right to get married for another 40 years. Without the right to marriage, same-sex couples are denied the same property rights, tax benefits, and immigration benefits that opposite-sex couples are granted by default. A woman in a lesbian relationship is doubly marginalized because she's not a man nor is she in a relationship with a man.

On an individual basis, queer people can also be denied inheritance when they are shunned by homophobic parents and therefore denied the opportunity to inherit generational wealth. Barely a single generation of women and queer people have had the same economic rights as straight men, even in parts of the world where some reforms have taken place. We must dismantle the idea that money is best left up to the straight, white boys and take control of our own wallets. Money affects everyone, and it's up to each of us to understand how.

Part of what makes debt so hard to grapple with is that it defies our intuitive sense of how money grows. Most of us are used to money being paid out at a constant rate, like a streaming service which costs $16 per month, or a job that pays $18 per hour, or $36,000 per year. These are all examples of constant, *linear* growth. Linear growth is steady, predictable, and makes a straight line when plotted on a graph.

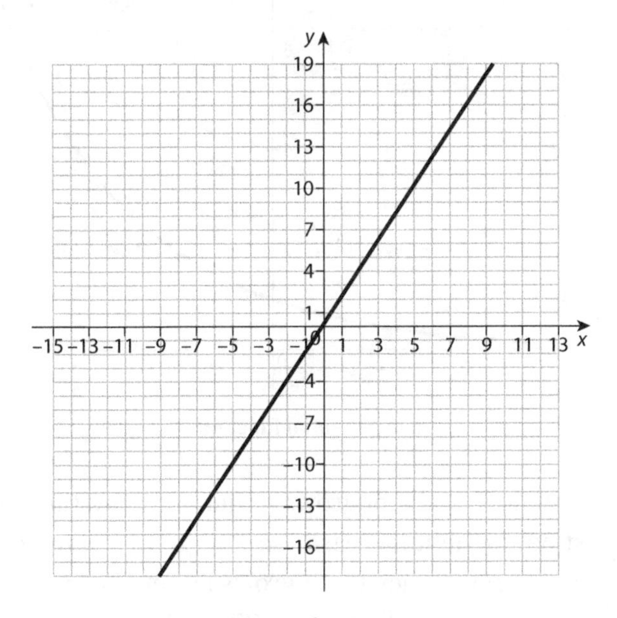

A linear function.

But the world is full of processes that don't grow in a linear fashion. Debt, for one, grows exponentially, and skyrockets (see page 136).

You can spot exponential growth in most things that grow by a percentage. This includes money matters such as compound interest, which stems from the notion of the time value of money that my dad was so insistent about. Suppose you take a loan out for $1, and it accumulates compound interest at a rate of 1% per day. In a day, your debt will increase to $1.01. In 30 days, it will become $1.35. After a year, your debt will become $37.78. At that point, 1% interest on $37.78 will be 38 cents, so the debt will grow at 38 cents a day instead of the original

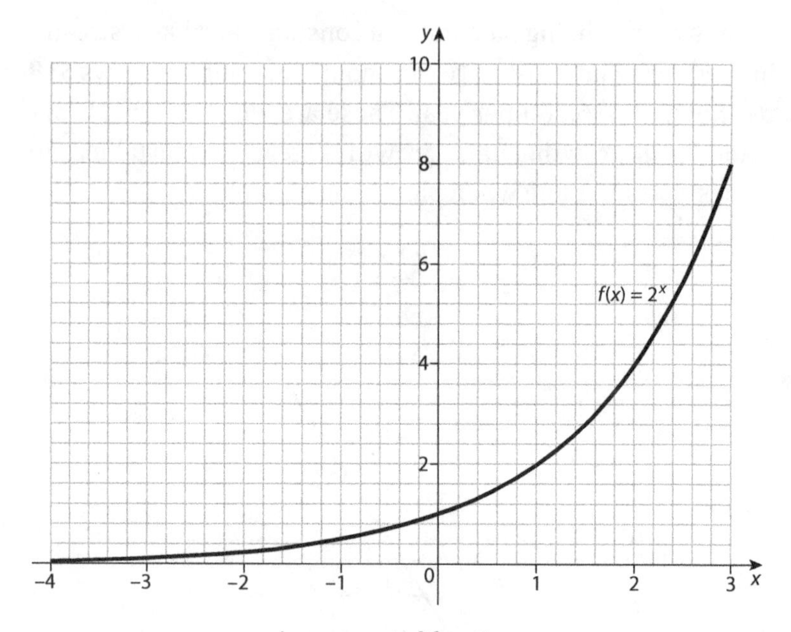

An exponential function.

1 cent a day. Since the debt grows larger each day, the amount of interest on the debt increases as well. In this way, not only does the debt grow larger, but it grows larger *faster*.

After 2 years you will owe $1,427.59. This is easily calculated by our equation $(1.01)^n$. The growth rate is 1%, or 0.01, so we calculate it by multiplying by 1.01, with n as the number of days. After 10 years, or 3,650 days, that $1 debt will have grown to $5,929,448,572,069,177.90, or $5.9 quadrillion! The small number 1% is starting to look pretty violent because 1% of $5.9 quadrillion is $59 trillion, which is the amount of interest that will be added the subsequent day. This is the result of slow, steady growth at 1% per day.

How is it possible that such a small rate of growth can turn a single dollar into a massive sum of money larger than the combined annual GDP (gross domestic product) of all the countries in the world in only 10 years? The answer is exponential growth. In real life, interest rates are far smaller than 1% per day (rates in real life are more like 1% per year), which is why there aren't any quadrillionaires (yet). But in principle, this is the same math that governs mortgage payments, student loans, and credit card debt. The key takeaway here is that *timing* is everything. If you have time on your side, you can take advantage of compound interest when you're young and grow a small investment into a large sum of cash by the time you retire. On the other hand, when accumulating debt, time is working against us; interest rates that appear small will grow so fast that they can eventually outpace the monthly payments a family can afford, putting them at risk of entering a cycle of endless debt.

Money might also be hard for people to manage sometimes because numbers on a screen are less tangible than physical bills. So to give a more concrete example, let's talk about the *real* issues of the world, like folding a sheet of paper in half.

The Magic of Doubling

I heard somewhere that folding a piece of paper in half more than 8 times is impossible. This didn't seem reasonable, so I decided to test the claim in an experiment video. As it turns out, I could only fold it 6 times before it became a centimeter thick and I couldn't physically fold it any further. Don't believe me? Try it! I'll wait.

Even if you could pull off a few more folds than me, the process would eventually feel like trying to fold a complete English-language dictionary in half. But imagine that somehow you could continue folding.

> **Riddle:** How thick would a sheet of paper become if you could fold it in half 42 times?
>
> **Solution:** The thickness would touch the moon.

This is one of those math facts that people find completely ludicrous when they first encounter it. Maybe that's why it was my very first video to go viral. Some people insisted that 2^{42} should actually be 84. But 84 is the answer to 2 *times* 42. However, 2 *to the power of* 42 is very different; it's greater than 4 trillion.

Whether you were willing to trust your friendly neighborhood drag queen or decided to reproduce the experiment yourself, you're probably still wondering how this could possibly be true. A notebook that has 300 sheets of paper is thin enough to fit in my backpack, so 42 folds seems like no big deal. But 42 *folds* gives you a much thicker end result than 42 sheets of paper. Each fold cuts the surface of the paper in half, and doubles the thickness. Remember that "fold" in this context means we have to fold the entire sheet in half, bringing one edge to the opposite edge. A small dog-ear fold like the one you do to the corner of a page in a book doesn't count, and neither does folding it into thirds as one might do when sending a letter in an envelope.

Folding a sheet of paper in half once will give you the thickness of 2 sheets of paper. Folding it twice will give you the thickness of 4 sheets $(2+2)$, and folding it a third time gives you a thickness of 8 sheets $(4+4)$. These are all the powers of 2. With each consecutive fold, the thickness of our paper doubles and follows this sequence: 2, 4, 8, 16, 32, 64, 128, 256, 512, 1024, 2048, and so on, with each term being twice as large as the previous one. To find the 42nd term in this sequence, which is the thickness of the paper after 42 folds, we simply have to multiply by 2 a total of 42 times. And 2^{42} gives 4,398,046,511,104 sheets of paper. If each individual sheet of paper is 0.1 mm thick, like common printer paper, then $0.1\text{mm} \times 4{,}398{,}046{,}511{,}104$ gives us a thickness of about 439,805 kilometers, enough to not just reach the moon, but overshoot it!

Unsurprisingly, no one has managed 42 folds in real life, but in 2002, a high schooler named Britney Gallivan broke the world record when she folded a sheet of paper in half 12 times, debunking the persistent myth that the limit was 8. She sought out the thinnest material that could still be called paper: a roll of toilet tissue 1.2 kilometers long. After the 12th fold, the paper was about a meter long and half a meter thick. (That's about 20 inches thick, for the metric-system resisters out there!) What's even more impressive, though, is that she came up with the equation explaining the math behind paper folding. If you want to fold a piece of paper n times with a starting thickness of t, Gallivan's equation tells you the minimum length L of paper you need to start with:

$$L=\left(\frac{\pi t}{6}\right)(2^n+4)(2^n-1)$$

I bet you didn't expect to see π here!

Technically, the equation tells you how much paper you *lose* in the folding process. The math behind the phenomenon is more about the ratio between the thickness and the length of the paper. Let me illustrate it with a diagram (see page 140).

Every fold causes some paper to be lost in the semicircles (shown with dotted lines). In the first fold, we lose a length that's equal to the semicircle's perimeter. A circle with radius t has a perimeter (also known as circumference) of $2\pi t$, and since the length of the dotted line is half a circle, the length is thus only πt. The second fold loses a length of another πt plus the perimeter of the larger semicircle, which is $2\pi t$. By the second fold, the total length lost is thus $\pi t+\pi t+2\pi t$. Gallivan's equation is a neat way of totaling the length lost after the nth fold, and thus represents the absolute minimum length of paper

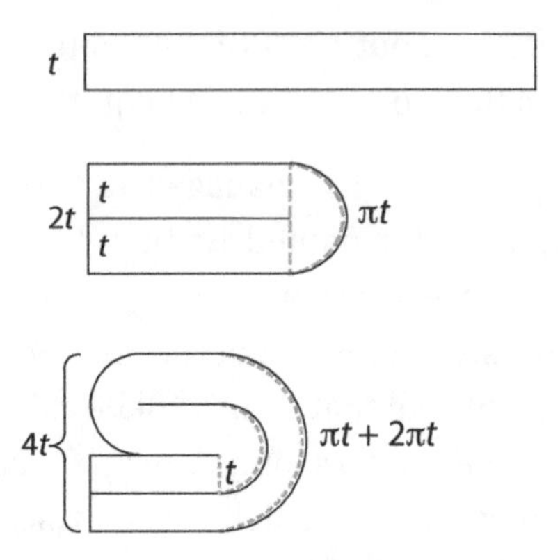

With each consecutive fold, some of the paper is being used up in the semicircle of the fold.

one needs to start out with. The equation also tells you when further folding is impossible: once the thickness-to-length ratio reaches 1 to π, you cannot fold any further because there won't be enough paper to make the semicircle that allows one edge to touch the other.

If we do the math with Gallivan's record-breaking 12 folds, we can see that the minimum length she needed was about 880 meters, which is longer than the height of Dubai's Burj Khalifa, the world's tallest skyscraper! But she actually started her experiment with 1,200 meters of paper instead, presumably to give herself some wiggle room. If she had wanted to try a 13th fold, she would have needed to start with a roll of toilet paper four times longer, or 3,520 meters. If she wanted to make a 14th fold, she would have needed a paper 14 kilometers long—longer than Mount Everest is high. Probably still doable, if all the world's drag queens, mathematicians, teachers, and students organized a folding party.

But 42 folds is hopeless. In order to fold enough paper to reach the moon, Gallivan's equation tells us that we would have to start with a sheet that's at least 10^{18} km long. To illustrate how long that is: if you stood at one end of this roll of paper and turned on a flashlight, it would take over 100,000 years for the light to reach the other end. This is the incredible power of exponential growth.

Exponential growth can be chaotic and surprising, like the spread of a viral video. If you or anyone you know has posted one, you may recall that at first the growing number of views was slow and steady, maybe reaching a hundred over the course of a week, then views started climbing by thousands at a time before exploding into the millions.

Riddle: My most popular video of all time is about a Möbius strip. Suppose that the first person who viewed the video immediately shared it with two others, and within a day each of those two people watched it and shared it with two new people, who each then took a day to view it before sending it to two others. If this pattern continues, how many views will the video have after 23 days?

Solution: We can draw the first few phases using a tree, starting with the first viewer at the top, the next two friends below them, each of their two friends below them, and so on (see page 142).

The first tier with one person can represent day 0. Each day, the tree grows larger, with 2 new viewers being added after day 1, 4 viewers being added after day 2, 8 after day 3, and 16 after day 4. These are all powers of 2! We can summarize all this information in a table (see page 142), showing the number of *new* views and the *total* number of views so far.

Following this pattern, the number of new views on day 23 will be 2^{23}, which is a huge amount. But what we're actually after is the number of *total* views by day 23.

There's a pattern in the number of total views if you look carefully. Notice that after each day, the number of

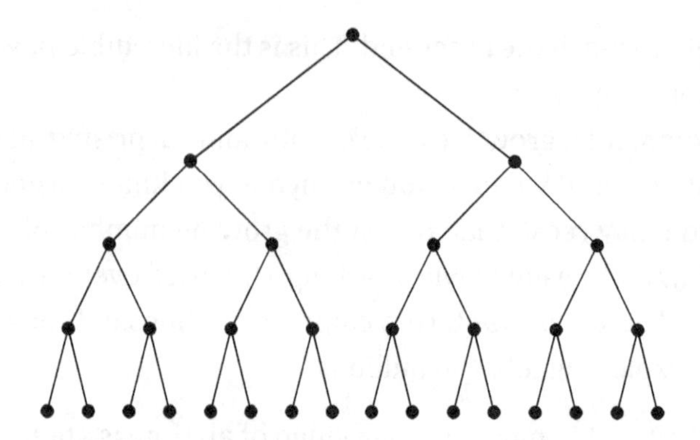

A tree showing exponential growth when each viewer shares a video with two new people each day. On day 4, there are 16 (2^4) *new* viewers.

The number of new views (and total views) after each day

Days passed	New views	Total views
0	1 (2^0)	1
1	2 (2^1)	3
2	4 (2^2)	7
3	8 (2^3)	15
4	16 (2^4)	31

total views is always one less than the next day's number of new views. On day 3, there were a total of 15 views on the video, and the next day 16 new views piled on. On day 4, the total views came to 31, and then the next day, 32 new views were added.

Every day brings more viewers to the video than have ever seen it before! This means that on day 23, the number of total views will be one less than the number of new viewers on day 24, which can be calculated by 2^{24} minus 1, which is equal to 16,777,215—this is greater than the entire population of Belgium!

But what if we were to extend this experiment for longer? If people continued sharing this viral video to two new people each time, then by day 31, it will have reached 4.2 billion people, and after 2 months it will have reached $2^{61} - 1$ views, which is the equivalent of all 8 billion people on Earth watching the video over 288 million times each. (Take that, "Gangnam Style"!)

Ideas spread exponentially, too. All it takes for an idea to make its way around the world is for one person to share it with two or more others. If it's a compelling idea, they'll share it with two more, and those two will share it further. In the beginning, it may feel like it's taking forever for the idea to make it outside of your bubble, but if you persist long enough, it may just take off!

RuPaul's Drag Race started as a concept dreamed up by three friends in 1993. Their pitch was rejected by several major TV networks, and they struggled for a long time to get the show off the ground. In fact, it took until 2009 to get their first season aired! Fast-forward to 2023: *Drag Race* has spinoffs in 14 countries—Thailand, the United Kingdom, Canada (hi!), Holland, Down Under, España, Italia, France, the Philippines, Belgique, Sverige, México, Brasil, and Germany. And there are plans to expand to South Korea, India, Singapore, Japan, and more! That's without even mentioning *RuPaul's Drag Race All Stars, UK vs. the World, Canada vs. the World, Secret Celebrity Drag Race, Queen of the Universe*, or *The Switch*.

Drag exploded from underground bars into people's living rooms. An idea, when nurtured carefully and pursued passionately, may take years to bloom, but it has the potential to reach the moon. If you feel like your life has hit a plateau where you've spent years stuck in the same place, consider the possibility that you're just in the slow, early stages of an exponential growth process. Because of its sneaky pattern of slow, sustained growth in the beginning, followed by rapid increases later on, it's hard to identify when something is growing exponentially, even when it's happening right before our false eyelashes.

Doubling Down

Hidden exponential growth can have negative consequences, as we saw with the frightening escalation of an initially small amount of debt. These consequences range from the personal to the planetary. Current estimates suggest our world population is growing exponentially at a rate of about 1.1% per year. After we had been hanging around the planet for about 200,000 years, the human population reached 1 billion around the year 1804. If the population had undergone *linear* growth, then it would have taken another 200,000 years to double to 2 billion. Instead, it only took 124 years. Less than a hundred years after that, we are 8 billion and climbing.

The global population is currently increasing at an annual rate of between 0.88% and 1.1%. But what does that really

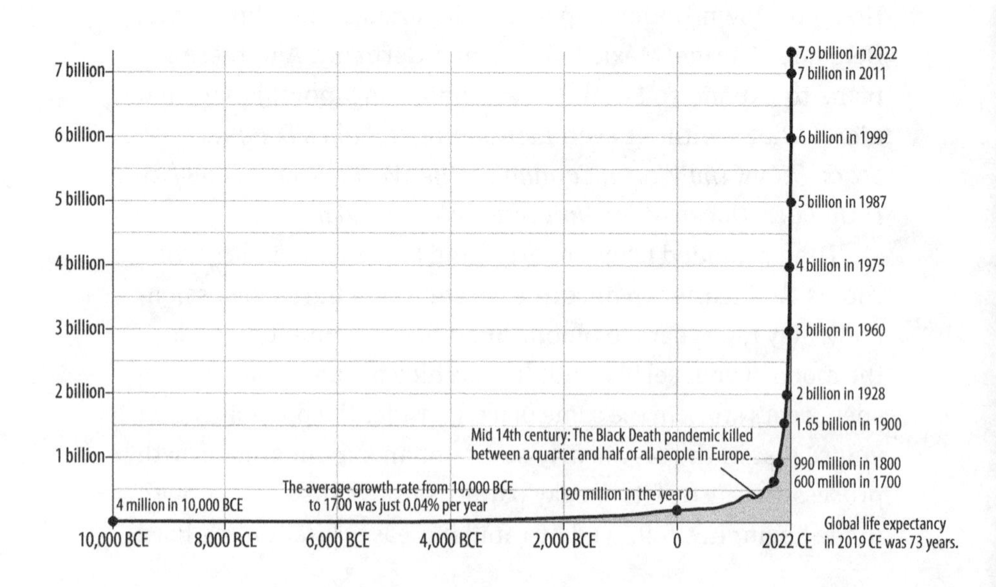

mean, and should we be concerned about it? Percentage growth rates can be confusing, especially if you don't know the right formulas to work with.

A growth of 1.1% doesn't seem alarming, until you take a look at the graph and see how dramatically the population is shooting up. At 1.1% growth per year, the population can be expected to double about every 64 years, meaning that if current growth rates continue the population will double to 16 billion by the year 2086, and 32 billion by 2150. In this case, 64 years can be called the *population doubling time*, which is the length of time required for the population to double in size.

Given a growth rate of $R\%$, you can calculate the exact doubling time by solving for n (n is the number of time periods—years, months, hours, or whatever—used to express $R\%$) in the following equation:

$$\left(1+\frac{R}{100}\right)^{n}=2$$

To solve for n, we will use an innovation called the *logarithm*, which essentially undoes exponentiation in the same way that division undoes multiplication and subtraction undoes addition. By the rules of logarithms, this simplifies to:

$$n=\frac{\ln(2)}{\ln\left(1+\frac{R}{100}\right)}$$

(A full breakdown of this solution is shown in the appendix.)

This value gives us the exact doubling time, but it's a little too much math for an on-the-spot calculation. So instead there's an easy formula you can use as a heuristic: the *Rule of 70*. Whenever you see a percentage growth rate, such as $R\%$, simply divide 70 by R and the answer tells you roughly how long it will take growth to double.

If the world's population is growing at 1.1% per year, the doubling time is approximately 70/1.1, or 64 years. If you have a bank account that accumulates interest at a rate of 5% per year, it will take approximately 70/5 = 14 years to double in size. Three doubling times is all it takes for a population to grow 8 times larger—that's like Toronto growing to the size of Beijing—and 42 doubling times was all it took to fold a single sheet of paper into a stack that touches the moon! Whenever you see a growth rate expressed as a percentage, there is exponential growth at play, and you can use the Rule of 70 to translate percentage growth rates into doubling times as a way to understand the growth from another perspective.

Repeated doubling can become an extremely powerful force, but so can its opposite: repeated halving. This is the opposite of exponential growth, and it's called *exponential decay*. It works by repeatedly reducing a substance or population by the same proportion each time. We experience repeated halving when a population is reduced by 50% at each time interval, like a tournament where half of the players lose in each round.

Riddle: Suppose 8 billion people on Earth participated in a rock-paper-scissors tournament. We start by pairing up everybody and making them play a single round of the game. If there are any draws, players are instructed to keep playing until a winner is decided. The losers are disqualified, and then the winners are paired up with each other in the next round. If an odd number of players ever results, then somebody is picked at random to be disqualified, ensuring the number of players stays even. This process continues until one winner emerges. How many rounds of the tournament will be required to crown that winner?

Solution: Starting with all 8 billion(ish) people on the planet, round 1 will eliminate half of them, narrowing the group of players to 4 billion. Rounds 2, 3, 4, and 5 will cull the field to 2 billion, then 1 billion, then 500 million, then 250 million, respectively. By the fifth round, the number of players left standing will be smaller than the population of the United States. By round 13, the number of players will decrease to under a million. Round 19 winnows the competitive field down to 15,258 players, enough to fit in a single stadium. Round 26 can be played in a movie theatre, since there will only be 238 players, with a mere 119 players walking out victorious. Round 31 will narrow that small group down to only 3 players, and one of these poor folks, with no partner to challenge, will have to be eliminated at random after coming this far! The final showdown, which ends in crowning the World Champion of Rock-Paper-Scissors will take place in round 32. This means that the ultimate winner of a tournament pitting the entire population of Earth against each other only has to beat 32 people at rock-paper-scissors!

We've seen how exponential growth is characterized by slow growth in the beginning and a rapid rise later on, but exponential decay works the opposite way—the fastest change happens first, slowly petering out afterward. Seven billion players were eliminated from our exciting tournament in just the first 3 rounds, but it took an additional 29 rounds to narrow the remaining billion down to 1.

Exponential decay can also be used in finance. Sadly, not all investments grow; some of them lose value over time, with a notorious example being cars. You may have heard people say that new cars lose 10% of their value the moment they're driven off the lot. This is down to the action of exponential decay.

The exponential function is a powerful instrument for all kinds of applications. One that has been pivotal in the last few years is modeling the spread of infectious diseases. It may be easier to picture how this would work in your backyard, rather than across the entire human population of our planet, so let's keep our thought experiment small.

Riddle: Suppose that a deadly fungal disease has made its way into your garden, and it doubles in size every 24 hours. You have only 100 days until it spreads to your entire garden. On what day will the garden be 50% infected?

Solution: Remember that we're facing exponential growth here, not linear. When something grows *linearly*, it should grow to half its size by the time it reaches half of its lifetime, suggesting the answer is day 50 out of 100. But this deadly disease is actually growing *exponentially*, doubling its size every 24 hours, meaning the answer is actually day 99.

We can prove this by working backward from day 100. If 100% of the garden is infected by day 100 and it doubles every day, then it must have been 50% infected 24 hours earlier, on day 99.

Working backward some more, the garden will have been 25% infected on day 98, 12.5% infected on day 97, 6.25% infected on day 96, and 3.125% infected on day 95. That isn't good news for you, the gardener. If no one ever warned you to look for early signs of the deadly fungus, you may not have even noticed a problem until the garden was 3% infected on day 95. Most of us wouldn't think of 3% infection as a big problem. You would point to all the healthy plants as proof that there was nothing really wrong, and you might call any dissenting voices conspiracy theorists and panic mongers. Maybe on day 97, when the garden is 12.5% infected, you'll finally realize that you need to intervene, but by then you'll only have three days to take action before the fungus wipes out everything worth saving.

The deadliest thing about diseases that undergo exponential growth is the long, deceptive period of slow spread, during which it's easy to procrastinate and mull over a solution. By the time you see what's going on, it's too late.

As our population reproduces exponentially, so does our consumption of natural resources, like fossil fuels, drinkable water, precious metals, and land. And some resources aren't growing along with us. Once we use up all the oil, coal, and natural gas buried in the ground, they're gone forever. The same can be said for the earth's supply of clean drinking water or biodiversity. It's worth asking: When will we reach a point of no return? Well, that question is beyond the scope of this book. Scientists all over the world are working on making those predictions using statistics, probability analyses, and exponential functions, but the future depends on what we do today. Instead, I'll pose a separate question: How will we know if the point of no return is near?

Our worldwide demand for oil is about 100 million barrels per day, and it grows by about 1% every year. Using the Rule of 70, we can estimate that it will double in $70/1 = 70$ years; meaning, in 70 years from now, our grandchildren will be consuming 200 million barrels of oil every day, and 70 years from then, our grandchildren's grandchildren will need 400 million barrels of oil every day. At this rate we will be using 6.4 billion barrels of oil per day after only 6 doubling times. What will happen after 10 doubling times, or 20?

Things get even more troubling when you think about how our oil use is so disproportionately shared across the world. In 2020, the average Canadian used about as much energy as 200 Haitians, and Haiti is one of the most vulnerable countries in the world to climate disasters. How much time do we have left to fix things? Are we like the gardener who is only noticing the deadly fungus at day 97? Or is it even later than that? After the last ice age—10,000 years ago—approximately 10.6 billion hectares of the earth was covered in forests, grasslands, and shrubs. Today, only about 5.7 billion hectares remain. We have already consumed close to 50% of the world's forests and green spaces, replacing them with grazing land and crops. Most of that deforestation happened in the last century. Mathematically, we are at day 99, with only one doubling period left until we destroy 100% of the earth's forests.

In the 2018 Marvel movie *Avengers: Infinity War*, the superheroes band together to stop Thanos (for those that don't know him, think Barney the purple dinosaur . . . if Barney were an intergalactic ecoterrorist). Thanos's home planet was destroyed by overpopulation, and he wants to stop it from happening elsewhere . . . by killing half of all the living creatures in the universe. *Spoiler alert*! With a sassy snap of his fingers, half of life on Earth disintegrates into dust. According to Thanos, the universe is then "perfectly balanced, as all things should be." After the movie came out, some fans shared that they thought Thanos did the right thing, and that he was more of an ecowarrior than an ecoterrorist. What would happen if Thanos snapped his fingers in our reality? Would that clean our beaches, revive our forests, and give the earth enough time to recover?

In theory, reducing our population by half would be like going from day 99 back to day 98. But exponential growth dictates that this would only extend our survival by a single doubling time, after which we would end up right back where we started, before we cheered for an evil dinosaur killing half our friends, families, and puppies. So much for the effectiveness of ecoterrorism. Moving to a new planet wouldn't solve things either. Even if we had access to three new Earths—bringing our total from 1 planet to 4 planets—at the same rate of growth we would still run out of space after only 2 doubling periods.

If we want to avoid extinction, mathematics tells us that the solution will require a fundamental change in the way we live, and science tells us that this change is within our reach. Our leaders will have to make unpopular decisions—ones that will anger large industries like oil, agriculture, transportation, and fashion. What's more, these changes won't only affect the top 2%; they'll disrupt the daily lives of most people on the planet. Thankfully, there are exciting emerging technologies on the horizon that could usher in an era of sustainability. Some of them are already here, but have yet to see widespread usage due to lack of political will and popular demand.

Good thing that ideas spread exponentially, too! Whether it's a viral Möbius strip video, *RuPaul's Drag Race*, or a way to quickly transition to clean energy sources, you have the power to plant a seed that can change the world. The key is time. Time is what allows populations to grow exponentially, and also how *Drag Race* became an international sensation. By taking small, consistent actions over a long period of time, we can create ripple effects that lead to positive change. You can apply this to your own life, too, whether your goal is to grow a big investment portfolio or a big wardrobe (or both).

Along with patience, the secret to harnessing the power of exponential growth is understanding that the rate of growth must be proportional to the thing growing. If we're doing it right, as life's challenges get harder, the growth we experience will rise up to meet them.

CHAPTER 7

Illegal Math

As a proud queen, I've dealt with my fair share of bullying and harassment over the years. But never in my life have I received the volume of death threats and hate mail that I have since I started teaching math. As an educator and entertainer (or as I like to say, an edutainer), I'm often invited to schools, conferences, and corporate events to give performances and presentations, sometimes about math, and other times about my personal and professional journey. Because of the way I dress when I perform, I am frequently called a "pedophile," a "predator," a "freak show," and a "groomer." I'm told to stay away from children. School administrators who hire me to give presentations have been hassled and threatened, and brands and corporations that use me in their advertisements are accused of pushing the "LGBTQ agenda" onto the masses. I feel a lot of guilt when people who work with me become targets of hate.

Here are some sample comments I received on one of my videos that simply discusses a cool math phenomenon:

- "Let's talk about dragging queer queens down the road behind a vehicle."

- "You will die if you touch our kids we promise oath keepers it's time to pick up our guns."
- "There is a millstone with his name on it."
- "Pew-pew is how to deal with homosexual people."

The topic of the video that angered people so much was magic squares. What is a magic square, you ask? It's a square grid of numbers in which the rows, columns, and diagonals all "magically" add up to the same number. Who knew it was such a divisive subject!

The loudest and most vitriolic uproar about drag has emerged among the "think of the children!" crowd. During the time I was writing this book, dozens of anti-drag bills and measures were introduced by right-wing extremist politicians in more than 14 states in the US. Some of these passed, some were rejected, and some are still under debate. Many of them seek to outlaw any adult cabaret performance from taking place on public property, or in any place at all that could be seen by a child.[1] The purposefully broad language used in this type of legislation would ban not only obvious events, like drag queen story hours at libraries, or the presentations I give teaching math in schools, but could also make Pride marches illegal, as well as drag brunches. Sponsors of these bills almost universally argue that Pride events and drag performers are trying to recruit children to LGBTQ lifestyles. An overwrought religious leader in Tennessee, where the first of these bills was advanced, lamented: "As a Christian pastor, events like these leave me heartbroken for our city and bewildered that our beloved Jackson is not safe from the cultural insanity seeking to reverse and even ridicule the basic laws of nature."[2]

In my town, there aren't any gay bars to host drag performances, so we hold daytime drag brunches or drag dinners at queer-friendly restaurants, where people can enjoy a meal with a show. The audiences are full of families, couples, work parties, and groups of friends. It's our job to uplift the crowd by getting them laughing, singing, and having a good time. Children are welcome at these brunches at the discretion of their parents and guardians. While drag shows are historically known to be provocative and irreverent, the all-ages shows I perform in carefully avoid any explicit material that might be considered even slightly inappropriate for families. Instead, we perform "dangerous" songs like "YMCA" by The Village People or ABBA's classic "Dancing Queen."

Adults and children alike are mesmerized by our sequin gowns, organza boas, and feather headpieces that dust the ceilings as we sing, dance, and do magic tricks. Drag brunches have even provided another venue to help me get young people interested in math. I once received a letter from a 10-year-old and their father, who had come to my shows together and wanted to tell me that they had become big fans of my YouTube math lessons. They even sent an addressed return envelope with a stamp so I could send them my autograph. It was so cute! Ever since, I've especially looked forward to seeing their family at my brunches. Don't get me wrong; there are still drag performances that are best kept in adult-only spaces, in the same way that movies and TV shows can be age-restricted on a case-by-case basis. But although it may have started as an adult-only tradition, drag has seen an explosion in mainstream popularity, and a branch of it has evolved into entertainment that families can enjoy together on TV and social media, or at live events.

While anti-LGBTQ hate groups are trying to frame drag queens as though we are "forcing" our lifestyle onto impressionable children, they never level similar accusations at parents who bring their children to football games where players regularly sustain serious injuries, or to comic book conventions where some cosplayers dress provocatively, or on hunting trips where they are taught to use weapons meant to end a life.

Unlike firearms how-to lessons, no children have been harmed by witnessing my brunches, just as no children were harmed by seeing John Travolta in drag in *Hairspray*, or Robin Williams as Mrs. Doubtfire. Nor have children been scarred by seeing Elsa, Belle, and Snow White impersonators read stories at the library, or sing songs at their birthday parties.

Come to think of it, when I was young, I did attend events hosted by men in long frocks. From the time I could talk, they were insistent on inducting me into their lifestyle, inviting me into a small, enclosed room where I was instructed to confess my deepest secrets to them. Instead of teaching me to love myself, they made me feel like I had sinned just by being me. The price for failure to conform to their lifestyle, they threatened, was an eternity of unimaginably painful torture in a terrifying place called hell. But you know what? I turned out fine anyway. Which must have really bothered them, because now those men in frocks are trying to render me illegal.

I can confidently say that none of us are trying to turn anyone queer, which is more than can be said for all the influences in my life that have tried to turn me straight. My only message is that math is fun and for everyone, no matter your gender, race, age, abilities, or dress size!

Math is at the heart of both my happiness and my success. In times of political and social turmoil, turning to numbers can be comforting. We like to consider numbers to be unwavering, unbiased, objective beacons of truth. If you can bring numbers into an argument, you are appealing to logic and reason, rather than sympathy, love, or kindness. It's a common belief that there's no arguing with numbers. But make no mistake, numbers can be controversial—perhaps even illegal.

Numerical Outlaws

Some numbers are considered unlucky, and others outright evil. You won't find a 13th floor in most hotels across Canada, and in Great Britain laws prevent drivers from putting the devil's number 666 on their license plates. On the other hand, while Chinese superstition sees 666 as a lucky number, you won't find a button for the unlucky 4 in any elevator in China. That's because the number 6 in Chinese is pronounced similarly to the phrase for "smooth" or "well-off," whereas the word for number 4 sounds a lot like the word for "death." Some numbers have even been censored. The string of digits 6/4/1989, the date of the Tiananmen Square Massacre, has purportedly been banned on Chinese search engines. A school district in Colorado prohibits students from wearing any jerseys featuring the number 18, due to its association with a gang that operates in the region.

Some numbers—like your credit card number, PIN code, or social security number—are illegal to distribute because of the information they encode. Anything you can see or share on a screen is represented by numbers in a computer. Every song you've streamed, textbook you've downloaded, and movie you've played breaks down to 0s and 1s. Pirating a movie is essentially the same as stealing the binary number which encodes the movie's information.

Back in the days of DVDs and Blu-rays, discs were encoded with a software that prevented users from making copies. In 2006, members of the online forum Doom9 made a breakthrough in pirating movies that used a particular encryption technique. Turns out, all it took was a secret numerical key that allowed users to bypass the encryption to watch and share movies they hadn't paid for. In response, the movie industry demanded the Doom9 website remove all mentions of the number. Unfortunately for the execs, their legal efforts had the reverse effect, creating a brand new celebrity number. People were shocked that corporations had attempted to make it illegal to distribute a number, so they responded by putting it on shirts, flags, and even tattoos. None of them were arrested or charged. But as an extra safety precaution, rather than sharing the *exact* number with you, here's the number that comes right after it: 13,256,278,887,989,457,651,018,865,901,401, 704,641.

Numbers can't *really* be illegal; it's only the information they represent that arouses controversy. I think we can all agree that math isn't an inherently bad thing, and it's obviously important, yet the message I get from lots of followers is that they hated learning math in school. Why is math so unpopular? I don't think it's just because of its difficulty, nor do I blame teachers. After all, writing essays is challenging, too, and so is memorizing the periodic table. I think the thing that makes math such a unique source of frustration is its humongous set of rules, which can feel strict, arbitrary, and confusing. *Don't multiply brackets without distributing first. When dividing by a variable with an exponent, always turn the exponent negative. Do not divide by zero, do not take the square root of a negative number, go directly to jail, do not pass go, and do not collect $200.*

Playing by the Rules

To most students, math seems like nothing *but* rules, all for solving puzzles that have little relevance to real life. To make matters worse, we rarely get any explanation for why these rules exist; we're just commanded to memorize them or else live with a lousy grade. If you're good at memorizing these rules, and also good at working under pressure, you'll be rewarded with high marks and maybe win a scholarship of your own. But many students, after failing classes or being treated like they're stupid for not understanding the rules, run away from math forever when they gratefully put their last exam behind them.

If you cringe when you think about math, or have cried over a tragic math grade, it isn't something to be ashamed of! You're far from alone. Math trauma is real! It's an actual term educators use to describe the consequences people can carry throughout their lives after they experience a tough time with math in school. I've done a whole episode about it on my podcast, "Think Queen." Math trauma can limit people's choices about the kind of future they want to build, damaging self-worth and derailing careers. What's more, it disproportionately affects people whose gender, race, language, or socioeconomic status doesn't line up with outdated ideas about who is likely to be "good at math." None of this is fair, and especially because *the rules aren't always the rules.*

Take this simple math problem, which went viral online exactly because people couldn't agree on what rules to apply to it.

$$6 \div 2\ (1 + 2)$$

I asked my followers on TikTok to vote on the solution to this seemingly simple problem, and it was a 60–40 split. Approximately 34,000 people voted that the answer is 9, and 23,000 voted for 1. Did one group forget to do their math homework?

When dealing with a problem that involves multiple operations like division, multiplication, and addition, students are sometimes taught a mnemonic like BEDMAS (or PEMDAS, BODMAS, or BIDMAS) to remember the rules about the order the operations must be performed in. Brackets, exponents, division, multiplication, addition, then subtraction. (As a sidenote, the *B* for brackets in BEDMAS and the *P* for parentheses in PEMDAS are referring to the same thing.) Playing by these rules, most of us can agree that the first step is to solve the $1+2$ inside the parentheses:

$$6 \div 2\,(1+2) = 6 \div 2(3)$$

So which step is next? If you follow BEDMAS, it's division. But what do you divide 6 by? Do you divide 6 by 2, and then multiply by 3 to get an answer of 9? Or do you divide 6 by the *product* of 2 and 3 to get an answer of 1? Think you know? Are you sure? Because the thing is . . . division doesn't *necessarily* go before multiplication. In fact, the *D* (division) and *M* (multiplication) in BEDMAS could just as easily be reversed. That's why some students learn the equally valid acronym PEMDAS, where multiplication is completed before division. In reality, multiplication and division should be seen as having equal priority, so the answer is to evaluate from left to right. The same goes for addition and subtraction.

But this still doesn't answer our problem. How are we meant to interpret $6 \div 2(3)$? Some of my viewers wrote to me arguing that you have to multiply 2 and 3 first because of the brackets (or parentheses) around the 3. That would mean that $2(3)$ is effectively the same as (2×3). On the other hand, if $2(3)$ is the same as 2×3 without any brackets, then the problem becomes

$$6 \div 2 \times 3$$

Which, evaluated left to right, gives 9. So the problem boils down to whether we interpret $2(3)$ as the same as either (2×3) or 2×3, a dilemma for which our teachers never prepared us when they taught us the rules of BEDMAS.

There is no objectively correct way to break this tie, so mathematicians have come up with generally accepted conventions. Historically, the division sign ($\div$) meant that one had to divide everything to the left of the $\div$ by everything to the right of it, which would indicate that $6 \div 2 \times 3$ is equal to $6 \div 6$, or 1. However, this convention is now about a hundred years old. Modern math textbooks suggest simply performing the division and multiplication from left to right, indicating that the correct answer to $6 \div 2(1+2)$ is 9.

In math, as in life, one way to resolve an ambiguous problem is to make the problem less ambiguous. When the division is written as a fraction with a clear numerator and denominator, then the order of operations becomes much clearer.

$$\frac{6}{2(3)}$$

$$\text{or } \frac{6}{2} \times 3$$

My conclusion is that this is a poorly written math problem because it can mean two different things. A better way to approach this problem would have been to make it clearer from the beginning, using fractions with an obvious numerator and denominator.

This problem went viral because it caused debate and barbs like, "People who answered incorrectly don't understand math!" While I love it when fun math problems incite debate and discussion, I don't like it when the debate is due to the ambiguity of the problem's phrasing because this can all too easily contribute to the myth that some people are just "bad at math." Mathematical rules may seem challenging when they are taught out of context, but they exist to give us a solid foundation. It's important that a mathematical sentence means the same thing to anyone reading it because that equation is a way of encoding information, often about a real-life quantity.

One branch of math especially concerned with rules is algebra. This is typically the class that first introduces students to the fact that math uses letters as well as numbers. The aim of teaching algebra is to show students how to solve for an unknown value, x, deploying a large tool kit of rules for how to manipulate equations like $3x^2+2x+1=4$.

The word "algebra" was coined by the Persian mathematician al-Khwarizmi (whom we met in chapter 2 and whose name also gives us the word "algorithm"). The word "algebra" comes from an Arabic phrase that originally referred to the surgical procedure of setting broken bones; al-Khwarizmi used it to mean "reuniting broken parts." Algebra is a mathematical kind of detective work, where you solve for x by manipulating and reuniting broken up equations. Its practical applications are endless.

Riddle: Without heels, I am 180 cm tall. But when I wear heels, my height becomes 190 cm. How high are my heels?

Solution: My heels must be 10 cm tall, because 10 cm is the difference between 180 cm (my height without heels) and 190 cm (my height with heels). If you were able to work this out in your head, then you have just mentally solved for x in an algebraic equation. We can convert this word problem into the equation:

$$\text{Height without heels} + x = \text{Height with heels}$$
$$180\text{ cm} + x = 190\text{ cm}$$

The desired variable x represents the height of my heels. We know intuitively that my heel height will be the *difference* between my height after putting on heels and my height before putting on heels. In math, that looks like a subtraction. Think of a mathematical equation like a pair of scales that are perfectly balanced around the equal sign; the value of the term(s) to the left of the equal sign equals the value of the term(s) to the right. As long as we adjust each side in the same way, the scales will remain balanced. If we want to manipulate one side by adding something to it (or subtracting something from it, or multiplying, dividing, etc.), we need to manipulate the other side in exactly the same way in order to keep the sides balanced. In this example, to solve for x, we can subtract 180 cm *from both sides of the equation.*

$$180 \text{ cm} + x - 180 \text{ cm} = 190 \text{ cm} - 180 \text{ cm}$$
$$x = 190 \text{ cm} - 180 \text{ cm}$$
$$x = 10 \text{ cm}$$

Converting a word problem into a mathematical equation may seem like it adds extra complications at first, but it can help us make sense of more complex problems. Give this one a try:

Riddle: I went shopping and bought some hair spray and lipstick. The total was \$36. If my lipstick cost double what the hair spray cost, how expensive was the hair spray?

Solution: To solve this riddle, let's use variables H to represent the cost of the hair spray, and L to represent the cost of the lipstick. Here's what we know:

The hair spray plus the lipstick cost \$36: $H + L = 36$

The lipstick is twice the cost of the hair spray: $L = 2H$

One of the rules you learn in algebra class is that we can substitute equations into one another. Since we know that L is equal to $2H$ (this is the second equation), we can substitute the value $2H$ into any equation in the place of L. So the first equation $H + L = 36$ is logically equivalent to $H + 2H = 36$.

$$H + 2H = 36$$

One H plus 2 Hs is equal to 3 Hs! Therefore . . .

$$3H=36$$

Finally, we want to know which number, when tripled, is equal to 36. We can solve that by simply dividing 36 into thirds:

$$H=\frac{36}{3}$$

$$H=12$$

Hence, the hair spray cost 12 dollars!

The basic laws of algebra are some of the most important math rules to understand because they are the foundations for many other crucial branches of math, from probability to statistics to game theory. And the more equations you solve, the more you get used to the rules.

But beware! There are some totally illegal activities! One of the most heinous transgressions is division by zero. This prohibited act can wreak all kinds of havoc in the real world, like crashing computers, but it also simply goes against the basic laws of logic and common sense. To see why dividing by zero is illegal math, consider this proof that claims $2=1$.

Let a and b be real numbers such that $a=b$.

(1) $a=b$
(2) $a^2=ab$ (multiply both sides by a)
(3) $a^2-b^2=ab-b^2$ (subtract b^2 from both sides)
(4) $(a+b)(a-b)=b(a-b)$ (factor out $a-b$ from both sides)
(5) $a+b=b$ (divide both sides by $a-b$)
(6) $a+a=a$ (replace b with a since they are equal by presumption)
(7) $2a=a$ ($a+a$ is equal to $2a$)

(8) $2=1$ (divide both sides by a)
(9) Therefore, $2=1$.

What foul sorcery is this? Of course, 2 can't really equal 1! At least not if we want numbers to be meaningful. In order to arrive at this obviously faulty conclusion, we must have committed a serious offense. See step 5, when we divided both sides of the equation by $(a-b)$. Remember that we started out by stating that a is equal to b, which means that a minus b must be equal to zero. So step 5 is a division by zero, which is against the rules! Dividing by zero leads to crazy logical contradictions like the assertion that 2 equals 1.

Say you want to question this basic law—out of curiosity or just because you're a rebel. Let's break down what division by zero really means. Take 20 divided by 0 as an example. Imagine you had 20 chocolate bars that you had to divide among 0 friends. How much chocolate does each friend get? 1 bar each? Infinite bars? None at all? Which answer makes more sense?

To think about it differently, let's divide the same way a mechanical calculator does, which is by repeated subtraction. For example, to divide 20 by 4, we simply count how many times we can subtract 4 from 20 before reaching zero. We start with $20-4=16$ (that's once), $16-4=12$ (that's twice), $12-4=8$ (three times), $8-4=4$ (four times), and $4-4=0$ (for a total of five times). However, if we divide 20 by 0, we can never reach zero through repeated subtraction. Even after subtracting 0 infinitely many times, we will be right where we started—at 20. Perhaps infinite chocolate bars is the better answer? It certainly sounds better!

As another example, think about what happens as you divide by smaller and smaller numbers while keeping the numerator the same:

$$\frac{100}{10} = 10$$
$$\frac{100}{1} = 100$$
$$\frac{100}{0.1} = 1{,}000$$
$$\frac{100}{0.01} = 10{,}000$$

You might notice that as the denominator (the number at the bottom of the fraction) gets closer to zero, the answer gets larger. But this does not indicate that dividing by zero means we get infinity. Look at what happens if we instead approach zero from the negative side:

$$\frac{100}{-10} = -10$$
$$\frac{100}{-1} = -100$$
$$\frac{100}{-0.1} = -1{,}000$$
$$\frac{100}{-0.01} = -10{,}000$$

As before, the denominator inches closer and closer to zero, but now the answer veers toward negative infinity. This is plotted as a graph on page 166.

So where is the solution when $x = 0$; that is, where does the curve cross the vertical axis? The curve approaches infinity from the right and negative infinity from the left. There's simply no good answer to $\frac{100}{0}$. Dividing by zero defies our logical framework. Division is simply a tool meant for numbers that aren't zero.

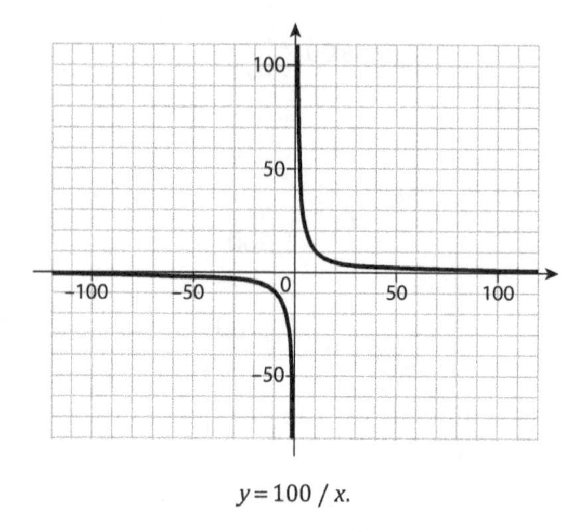

$y = 100 / x.$

Breaking the Rules

I know algebra can look daunting at first glance. If you aren't used to manipulating equations like this, it can seem like a complicated board game with endless rules and symbols, and plenty of illegal moves. When I was growing up, schoolmates often described themselves as "math people" or "art people," drawing a line between being good at numbers and logic and being good at creative thinking and social skills. I think that this establishes a limiting belief, which is especially harmful for young students. Your girl, Kyne, would stand up in any court of law as evidence that you can be both!

Plenty of people seem to think that math and drag are polar opposites of each other. While math is defined by rules, drag appears to abhor them. After all, drag is about subversion. Not so long ago cross-dressing and homosexual activity were illegal in many Western countries.

A gay subculture was born somewhere between the sixteenth and nineteenth centuries in the British Isles, where denizens developed their own slang to communicate with each other while hiding themselves from homophobic outsiders and undercover police. Many gay men joined the circus, cruise ships, or the merchant navy, and they disseminated their lingo around the world. This slang language was called Polari, and it included words like "camp" (meaning effeminate), "naff" (meaning bad or corny), "trade" (referring to a casual sexual partner), and perhaps even the word "drag" itself, referring to clothing that one typically isn't expected to wear.

Laws that were created to uphold public decency were used to enforce a normative gender binary. Where it wasn't outright illegal, police simply found other reasons to harass queer and gender nonconforming people. And that behavior certainly wasn't limited to one side of the Atlantic.

William Dorsey Swann was born into slavery in America around 1860. After the Civil War, his newly free parents bought a farm, and Swann took his first job as a hotel waiter. In the 1880s, he started organizing secret drag balls in Washington, DC, becoming the first person in recorded history to call himself a "Queen of Drag."

According to journalist and Swann scholar, Channing Gerard Joseph, most attendees were formerly enslaved Black men or the children of enslaved parents, who would show up—resplendent in silk and satin gowns—to dance and socialize.

On April 12, 1888, a police officer stationed in the northwest area of Washington, DC, watched three individuals dressed like "colored females" in flashy, low-neck and short-sleeve dresses step out of a carriage and disappear into an adjacent building. Thinking they looked "suspicious," the officer went into a nearby hospital and asked if he could peek through their window for a better view inside the house in question. The officer, spying 30 men, many dressed as women, wasted little time in calling for backup. It was the first time local law enforcement would raid one of Swann's illegal drag balls. While the other dancers tore off their dresses and wigs, leaping out windows and sprinting away, Swann was said to have "stood in an attitude of royal defiance," which led him to become the first person in the United States to be charged with the crime of female impersonation. "You is no gentlemen," Swann reportedly proclaimed as the officer laid hands on him. The following day, the *National Republican* (later merged into the *Washington Post*) ran a breathless article on the front page, " 'The Queen' Raided."[3]

Pearl-clutching journalistic coverage is the only reason we know the name William Dorsey Swann today.

Eight years later, Swann was arrested again, this time for "keeping a disorderly house," also known as running a brothel. He fought the charge, escalating his petition for a pardon all the way to President Grover Cleveland. The request was denied, but Swann went down in history as the first American to pursue legal action to defend the LGBTQ community's right to gather. He retired from the drag scene around 1900; but his brother took over, and the ball tradition continued to evolve over the next century, spreading to cities like Baltimore, Atlanta, Philadelphia, and New York. After Swann died around Christmas of 1925, local officials burned his house down.

The institutionalized repression of queer people didn't abate much in the United States over the next half century. During the post–World War II Lavender Scare that ensnared mathematician John Nash, the FBI surveilled known homosexuals, their friends, and their favorite establishments, creating a culture of increased secrecy and fear. LGBTQ people were forced to find safe havens where they could freely express themselves. But the police would raid these locations, too, lining people up to check that their sex matched the clothes they were wearing, arresting them for "degenerate disorderly conduct" or other offenses if they didn't like what they found. There was also the constant fear of entrapment; it was perfectly legal for police officers to pretend to be gay and try to pick up patrons. If they took the bait, they were arrested for breaking yet another law: "homosexual solicitation."

The Stonewall Inn on Christopher Street in New York City was one of these safe havens for queer people. It was a special favorite of drag queens, who were often shunned at other clubs that tolerated gay patrons. Since any gathering of gay people was illegal, the Genovese crime family, who owned the bar, bribed the police to look the other way, while making money by overcharging on drinks and blackmailing closeted patrons. But safety was far from guaranteed.

In the early morning of June 28, 1969, police raided the inn and, as usual, they lined up the patrons, arrested anyone in drag, assaulted people at will, and searched for bootlegged liquor. But something was different that day. Instead of cooperating with the police, the patrons fought back. Among them was the butch lesbian drag king Stormé DeLarverie, who may have been the one who shouted, "Why don't you guys do something!"[4]

While trans drag queen Marsha P. "Pay It No Mind" Johnson has become legendary as the person who started the riot by hurling a brick, other eyewitnesses say it was thrown by Zazu Nova, a trans sex worker and the self-described "Queen of Sex."[5]

The exact details of that night will probably remain a mystery, but what we know for sure is that a crowd of hundreds of working-class queer people formed outside, throwing pennies, rocks, bottles, and bricks. When the police locked themselves inside the bar, the rioters set it on fire. Those who consented to be interviewed in the aftermath stated that they felt they had nothing left to lose. The riots continued for three days. They marked a major turning point in queer activism.

Post-Stonewall, the regulations and laws that had effectively made it illegal to be queer began to disintegrate, and drag queens played a major role in making it so.

Today, Pride parades are held in increasing numbers at the end of June every year to commemorate the anniversary of the Stonewall riots, but our freedoms are still being threatened by politicians like Florida governor Ron DeSantis, US representative Marjorie Taylor Greene from Georgia, and Tennessee state representative Chris Todd. Anti-LGBTQ extremists carry on the crusade against queer people, regurgitating the same, centuries-old arguments about social morals and public decency. We are arbitrarily labeled "groomers" and "perverts" because these extremists want impassioned mobs to forcibly remove us from our events and push us back into the closet (or at best, into dark, illegal, underground clubs), whether through violence or public shaming. Some states, provinces, and even whole countries may not allow this math book you hold in your hands to be read or distributed to minors because it will be seen as LGBTQ propaganda.

Expressing yourself by dressing the way you please is still revolutionary.

Anyone who tries to make drag illegal is missing the point entirely. The suppression of queer people is precisely what makes the LGBTQ community strong. Our shared trauma brings us closer, whether we experienced bullying inflicted by peers as children, torment from some of our parents, rejection from our churches, oppression from our governments, or all the above. But our culture will never disappear because, even after campaigns of violence and terror, there are queer people in all corners of the world, from the Philippines, Uganda, Saudi Arabia, Iran, and Russia, to Tennessee, Texas, and Florida. I, for one, am *done* with living underneath the floorboards of society where homophobes have tried to push us. In many parts of the world, we're making strides toward freedom, equality, and visibility.

Right now drag artists are freer and more visible than we've ever been. If it feels like queer people are everywhere these days, it's not because of any agenda; it's only because your previous sample size was biased. Queer people have always been around. You can see us at school bake sales, running for office (and winning), electrifying crowds of sports fans, and performing in world-class theatres. Drag, despite being controversial, has become an international phenomenon.

The Rules of Drag

Given the long history of drag queens fighting against repressive laws, it may surprise you to hear that, while rebellious, drag itself is not lawless! In fact, it is circumscribed by many rules, so in that way, it's a lot more like algebra than you might think!

What do drag laws look like? Well, commandment number one is knowing your lyrics. Whether you're lip-syncing or singing live—but especially if you're lip-syncing—you absolutely, positively, no-exceptions-allowed, must know your lyrics.

As fans of *RuPaul's Drag Race* know, another hard and fast rule is that a queen should never remove her wig (unless there's something fabulous to be revealed underneath).

Drag pageants are even more ruthless. In the pageant world, contestants are ranked on minutiae, like the height of their heels or the convexity of their hips. When competing for Miss Gay America, for instance, your evening gown must end a half inch above the floor, unless of course you've chosen to wear a ball gown, which should touch the floor. Any ill-fitting gowns, buttons out of place, loose threads, or visible tattoos will deduct from your score, and so will submitting competition music on an unlabeled CD, showing poor sportsmanship, or tardiness. Extravagant evening gowns must be paired with simple jewelry so as not to overpower the look. Your sleeve should be no shorter than your wrist and fall no longer than an inch past it, or else points will be deducted. Points will also be deducted if your shoes are scuffed, so you better make them look perfect even if they'll be covered by your dress, just in case the judges catch the smallest glimpse of them as you walk up the stairs.

The category with the most points up for grabs is the talent portion, where contestants usually lip sync or sing live. And faithful adherence to commandment number one is not enough. Are you finishing each word crisply, and emoting to the song? If you're performing with backup dancers, can you keep up with them, or are they doing all the work?

At the end of each pageant, you can collect your scoresheet and see your marks, just like a student collects their math test to see what they did right and what they did wrong, so they can try to avoid making the same mistakes again. Pageant-winning drag takes years of experience and discipline, and lots and lots of coins. Good drag might look effortless, but it takes practice. The illusion of effortless mastery threads through our ideas about talented mathematicians, too. Celebrated mathematicians are traditionally portrayed as prodigies to whom math comes naturally. Think of movies like *A Beautiful Mind* or *Good Will Hunting*. This stereotype of the magical unicorn math genius contributes to the notion that one must be naturally gifted in order to excel at math. But good math, like good drag, is a skill that requires time and patience. With enough practice, you can become well-versed in all the complicated rules of algebra and pageantry, and you will find that those "weird" rules become more intuitive and easier to apply.

But what happened to drag breaking all the rules?

Drag is a form of art, and art has few rules, but sometimes constraints can actually spark creativity, like how a haiku is restricted to three lines with a 5-7-5 syllable structure, or a Shakespearean play is written in iambic pentameter, which stresses every other syllable: The *queen* who *plays* the *flute* will *steal* the *show*.

In ancient India, all mathematical texts were written in Sanskrit verse, which followed a specific structure and meter.[6] Mathematicians were effectively writing poetry and math at the same time, which doubled the constraints on their work. If a mathematician wanted to write the word "two," for instance, but it didn't fit into the meter of the poem, they might have used the word "hand," "eye," "foot," or anything that comes in pairs. Instead of using the word "add" in places where it didn't fit the rules of the poetry, the mathematician Aryabhata used synonyms like "mix," "heap," "accumulate," "unite," or "yoke." This made the work more complex, but it also made it easier for people to remember and pass on orally, ultimately creating a more cohesive, beautiful, and impactful literature of mathematics.

Queens work under the constraints of the archetypal act that most expect to see from a drag show: there's song and dance, an outrageous outfit exaggerating gender expression, comedy, and some tongue-in-cheek fun. It's these constraints that create innovation. To get the look of a curvy body, many drag queens resort to robbing couches of their foam to pad out our hips. To get the look of a rounder, more feminine hairline, some performers use colored hair spray or a lace front wig. To get the look of higher, thinner brows, we glue down our natural ones and cover them up with makeup. I've seen performers use construction paper to make cartoony eyelashes, mannequins attached to poles as moving backup dancers, and, once, a can of hair spray to imitate magic ice powers in an Elsa *Frozen* number (shoutout to the Filipino drag queen Katkat). The creativity I've seen from drag queens could put rich pop stars to shame!

You might find it ironic that drag has so many "must-do's" when it ought to be about being subversive, but to me the constraints are what make drag entertaining and rewarding. There's something about a good old-fashioned pageant queen that's just—*chef's kiss*—correct, and that's because they studied the scoresheet.

And yet, every now and then, someone successfully pushes past the agreed-upon boundaries to widen our perception of what is beautiful, or even possible. Some drag queens perform with their natural, full beards; others wear scary makeup instead of looking glamorous; and some perform songs traditionally sung by men instead of women, like Aerosmith's "Dude (Looks Like a Lady)," or James Brown's "It's a Man's Man's Man's World." I particularly love Morgan McMichaels's performance of Beyoncé's "If I Were a Boy," where she starts the song in full drag and lip-syncs while wiping off her makeup and changing out of her gown to finish the second half of the song in jeans and a T-shirt. Since drag queens already play with the boundaries of gender, I think it's especially brilliant when artists push it even further and add extra layers of queer chaos. You can mess up the lyrics to a song if you're doing it with intention and humor, and you can rip your wig off if the music compels you to; the audience may even cheer you on for it. You can wear flats instead of heels if your feet are hurting, and you're allowed to rock your natural body sans padding or corsetry. Many drag artists do not even identify as queens at all! Drag kings and drag things represent royalty of all genders—and honey, I am here for it!

The Rules are Flexible

Commenters on the BEDMAS problem (at the beginning of this chapter) who ranted that the US math education system is a failure because people think that $6 \div 2(1+2)$ is equal to 1 are thinking too small. There *are* shortcomings in math curriculums, but they don't stem from rules not being enforced stringently enough; on the contrary, students aren't learning that math rules have some elasticity to them! Even our sacred rule against division by zero has some flexibility.

Consider 0 divided by 0. How would this work? On one hand, you might remember the rule that a number divided by itself is always equal to 1, but on the other hand there's also the rule that 0 divided by any number is equal to 0. But on the third hand, dividing by zero is illegal math, and that rule trumps all. However, there *is* a mathematical way we can get around the rules without flouting them to make sense of a number like 0/0. The key is to do it with the care and finesse of a queen. If we break rules haphazardly, we end up with messy drag and messy math.

In order to carefully approach division by zero, mathematicians had to invent a whole new branch of math—calculus. Calculus deals with rates of change. Think of the speedometer in your car going up and down as you accelerate and brake. Imagine you're driving a car around town and onto the highway to reach another city. After an hour, you reach your destination 80 kilometers away from home. You can calculate your average speed over the journey by taking the distance traveled and dividing it by the time: 80 kilometers divided by 1 hour gives us an average speed of 80 km/h. However, during the journey you watched your speedometer rise and fall with every break and acceleration. As you merged onto the highway, you watched the speed climb up to 80 km/h, then 85 km/h, then 90 km/h, and as you approached stoplights you watched the speed quickly drop back down to 0.

We know that the car actually traveled a total distance of 80 km over a time period of 1 hour, so what do we make of all these other speeds in the middle? Your car's speedometer is actually measuring the *instantaneous speeds* of the car. This is the speed read at a single, isolated moment in time. But since speed is defined as distance divided by time, what does the speed at a single moment look like? It looks awfully close to calculating distance/time when time is equal to 0. But remember, we can't divide by zero. After all, how much distance can be traveled in 0 seconds anyway?

To avoid a division by zero, your car calculates distance/time using multiple, small intervals of time, like 1 second, then 0.1 seconds, and then 0.01 seconds. As the time interval gets shorter, the distance traveled will also get smaller, but the ratio of distance/time will start to approach a meaningful number, and this is the reading you see on the speedometer. In mathematics, we call this the *limit*. With limits, we can make some sense out of dividing by zero by evaluating equations using numbers that are infinitesimally small, and increasingly close to "the big 0."

Let me give another example which doesn't involve cars, but uses graphs instead. Take a look at the graphs of $y = \sin(x)$ and $y = x$.

Notice that both graphs equal 0 when x is equal to 0. Now what if we divided these two functions by each other, and graphed $y = \sin(x) / x$?

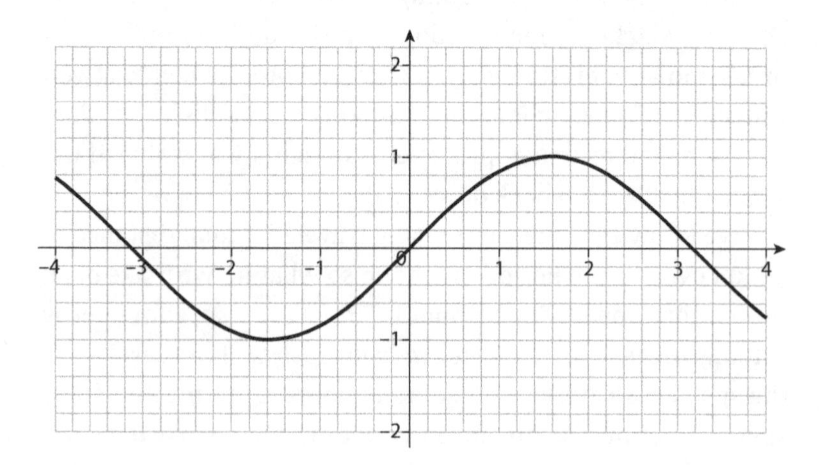

$y = \sin(x)$.

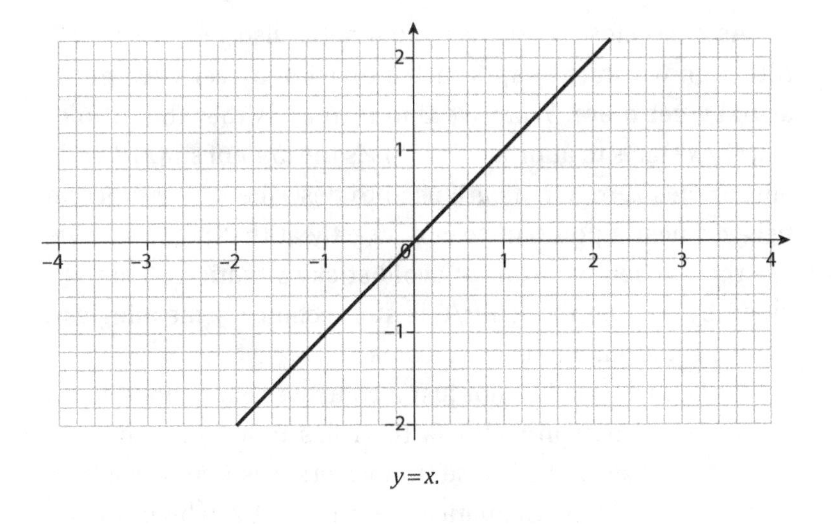

$y = x$.

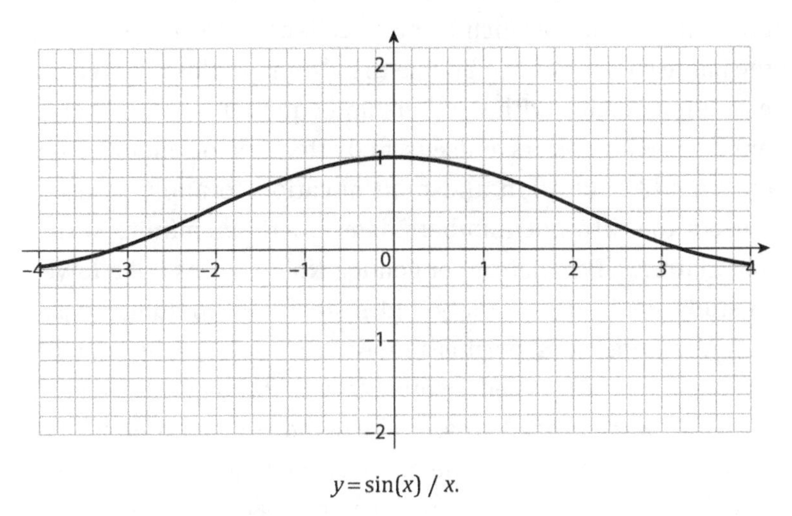

$y = \sin(x) / x$.

If we input $x = 0$ into this new equation, it would look like $\sin(0)/0$, which is 0 divided by 0. An absurd question! Yet this graph seems to have an answer, because at $x = 0$, the graph seems to suggest the y-value is 1!

Taking a closer look at this, we can examine what happens when x is just *close* to zero. When x is equal to 0.5, $\sin(0.5) / 0.5 = 0.959$. And when x is equal to 0.1, $\sin(0.1) / 0.1 = 0.998$.

As x gets closer to 0, the denominator also gets closer to 0. Although we start to approach a division by zero, we never actually get there. What we find instead is that the ratio of $\sin(x) / x$ starts to approach 1! This is the *limit* of $\sin(x) / x$ as x approaches zero. Limits are used to describe how a function behaves near a point, instead of exactly at that point, and we use them when we want to avoid sticky situations like 0 divided by 0. Even though 0/0 is still technically meaningless, we can get around it using calculus.

Sometimes it takes not just a controversial method but a controversial person to break the rules in a thoughtful and productive way. One of these trailblazers was Emmy Noether, a German Jewish mathematician born in 1882, who is among the founders of a branch of math called *abstract algebra*. As the name suggests, it's a bit like the algebra in this chapter, but instead of dealing with numbers and equations, abstract algebra deals with objects and sets, and attempts to generalize operations like addition and multiplication to apply to other objects that are not numbers, like functions, infinite series, and matrices. Because of her gender, Noether was not allowed to study math in university and instead had to audit classes with the professors' permission. After completing her doctorate in this unsanctioned fashion, it was still against the rules for her to work in academia, so she persevered without pay for years by lecturing under the name of the male mathematician David Hilbert. In 1933, she received a letter from the German government notifying her that it was illegal for her to work as a professor because she was Jewish, so she fled to America, where she accepted a position at Bryn Mawr, a women's college. Despite all those obstacles, she is considered to be among the greatest mathematicians of all time. How appropriate that she made ground-breaking discoveries in abstract algebra, a field that turns a lot of mathematical rules upside down—there are even situations where it may make sense to say that $0 = 1$.

Math, just like drag, can be a joyful dance that delights in applying rules while elegantly pushing boundaries. What we ultimately find is that by following rules, we discover ways to transcend them. There are no commandments that can't be bent or broken, and there are no illegal numbers, just as there should be no illegal people. Without breaking rules, we never would have discovered calculus, abstract algebra, infinity, or imaginary numbers. And without breaking laws, we wouldn't be achieving the liberation of millions of marginalized people across the world. We will continue to transcend the rules because every great discovery is born just over the boundary of what we think is possible.

CHAPTER 8

Queer Geometry

At 12 years old, I was a super-devout Catholic. We all go through phases, okay! I was so inspired by the sermons at our family's church that my internal monologue became a constant stream of conversation with God. During my Tumblr expeditions, if I happened to see any posts that cast God in a negative light, I would literally cover my eyes while I hurriedly scrolled past. I was so afraid that my feelings as a gay kid were a terrible sin; I even decided that I should be married to the church, just as priests and nuns were. But that all changed one day when I came across a book in our school library: *Cosmos*, by Carl Sagan.

I was fascinated by what I read about the big bang theory and the birth of the universe, and I immediately jumped down an internet rabbit hole of information about creationism, evolution, and the big bang. I landed on the other side dizzy, realizing that I couldn't be both a man of science and a man of God. (I was only a little boy, but this is really how I thought—I'm so dramatic like that.) I started to ask myself *why* I believed in God, and I couldn't formulate a convincing answer.

I stomped into my parents' room and asked them why we were Catholics, and why we believed in God and heaven and hell and other things like that from the Bible. They insisted that it was just tradition. For my family, going to church was as ingrained as eating rice for dinner or taking our shoes off when entering a house. It was the way they grew up, and how their parents grew up before that. But this reason wasn't convincing enough for me. So, I decided I was an atheist.

I know religion is a complicated and personal subject for a lot of people, and I don't advocate that everyone should go losing their traditions and faiths. I do, however, think that it's a good thing to question your own belief system once in a while, and really evaluate *why* you believe the things you do. Every one of us is programmed at an early age with a default mode of core morals and beliefs and, if we've outgrown them, we may not realize they're holding us back.

I once made a video talking about drag and queer people in education, and I received this comment: "Let kids have the real default way of thinking. There's never been anything other than two genders."

I had to pause for a moment.

What is the "default" way of thinking? If gender nonconforming people inhabit this planet along with everyone else, why wouldn't we be included in this person's de facto reality? We already know of other civilizations where nonbinary genders and same-sex relationships were seen as normal, like the Filipino *bayog*, and two-spirit Indigenous people in North America. This "default" mode that we hold to be obvious and self-evident isn't based on any objective truth but rather on social, political, religious, and cultural traditions, which are in constant flux.

Sometimes that flux provides space for more expansive views to emerge. But other times societal shifts can cause a person's—or a culture's—frame of mind to narrow, instead.

In 2022, Florida lawmakers passed the "Parental Rights in Education" bill, better known as the "Don't Say Gay" bill, which prohibits any classroom discussion of sexual orientation or gender identity from kindergarten through third grade. Proponents of the bill are trying to expand that rule through eighth grade, and prohibit students from changing their pronouns up until twelfth grade. The bill's most prominent fan, Florida governor Ron DeSantis, had this to say during a press conference leading up to the decision: "My goal is to educate kids on the subjects—math, reading, science—all the things that are so important. I don't want the schools to kind of be a playground for ideological disputes."[1] His press secretary was more transparent, tweeting that the "Don't Say Gay" bill is more accurately described as an "Anti-Grooming" bill.[2]

Personally, I think that schools should prepare young people for adulthood. While some argue that acknowledging their LGBTQ neighbors' presence on this planet is forcing them into some kind of radical "woke-ism," the truth is that their children are growing up in a diverse world. Students should understand that people in a free country choose to have different sorts of relationships and family dynamics, like two mothers, or a single father, or a nonbinary parent.

I've met teachers who have come out in their professional lives as gay, trans, or nonbinary, and I've given talks at schools that celebrate Pride month and allow students to identify with whatever pronouns they please. These are all practices and policies that fervent anti-LGBTQ parents and politicians would have you believe are youth "indoctrination." Whether or not Ron DeSantis finds these different identities and relationships legitimate or peculiar should have no impact on children's need to learn that all sorts of people are equally worthy of respect and kindness. At least that's what Jesus would say!

I agree with DeSantis on one thing, which is that schools should teach important subjects like math. And that means schools should teach students how math *really* operates. It can be provoking and eccentric, and it defies rigid frameworks. It encourages us to question authority, reassess our predetermined models of reality, and break away from tradition to blaze new paths that may appear anarchic or unsuitable for any practical application. Unfortunately, many students seem to be absorbing a completely opposite impression. Most think that math is all about unforgiving strictures and uniformity. DeSantis should be appalled that math has gotten such a bad reputation! One of the branches of math that has gotten the worst rap also happens to be one of my favorites and is—maybe not coincidentally—queer in all the best ways. It's that misunderstood marvel, geometry.

You might remember geometry being the class where you learned about circles, triangles, squares, hexagons, and formulas like πr^2 and the Pythagorean theorem.

The *Pythagorean theorem* states that if you start out with a right-angled triangle, then the squares of the two sides adjacent to the right angle will add up to be equal to the square of the *hypotenuse*, the side opposite the right angle. If we label the length of the hypotenuse as c, and the length of the other two sides a and b, then we can write this as $a^2 + b^2 = c^2$. The Pythagorean theorem is one of those classic examples of stuff we're forced to memorize in school that we never use in the real world.

But there's so much more to geometry than the Pythagorean theorem!

Geometry is the math of everything you can see. All around you are shapes, lines, structures, directions, and horizons that seem to extend forever. Geometry gives us a language to describe our world by measuring objects, calculating distances, and judging angles. It also allows us to extrapolate from what we see to build new things. With it, we can measure immense objects like the earth or the moon, even though they're many times larger than any ruler. That's because geometry allows us to draw big conclusions from little information. Geometry isn't about conformity or rigidity; it's about observing patterns and building great things from the very small.

The oldest text on geometry we have found to date, called the Rhind Papyrus, dates back to ancient Egypt around 1500 BCE. The text begins with the scribe introducing himself: "Accurate reckoning. The entrance into the knowledge of all existing things and all obscure secrets. . . . It is the scribe Ahmes who copies this writing." The text then goes on to read like a student's math homework, with a list of math problems and solutions.

See if you can try your hand at an ancient Egyptian math problem:

An estate consists of seven houses. Each house has seven cats, each cat kills seven mice, each mouse eats seven grains of barley, each grain would have produced seven hekats (units) of wheat. What is the sum of all of these?

Working it out, we start with 7 houses, 49 cats (7 for each house, and $7 \times 7 = 49$), 343 mice (7×49), 2,401 grains of barley (7×343), and 16,807 hekats of wheat ($7 \times 2{,}401$). These all add up to 19,607, which the scribe Ahmes writes.

Put more simply, we can write this problem with symbols:

$$7 + 7^2 + 7^3 + 7^4 + 7^5$$

Modern students recognize this as a *finite geometric series*, finite because there are only 5 terms to be added instead of infinitely many terms, and geometric because the terms grow by a common factor of 7 each time. There is a shortcut for adding up terms in a general equation like this one:

$$1 + r + r^2 + \ldots + r^n = \frac{1 - r^{n+1}}{1 - r}$$

In this case, since we want to start with 7 and not 1, we have to multiply both sides of the equation by 7. Then, we replace r with 7, as well, since 7 is the common factor, and replace n with

4, so that the last term added will be 7×7^4, which equals 7^5. We get:

$$7 \times \frac{1-7^5}{-6} = 7 \times 2{,}801 = 19{,}607$$

This shortcut might seem like extra unnecessary steps, but it's useful for calculating larger series. What's amazing is that thousands of years ago the scribe was already well aware of this shortcut! In the next column over, he writes 2,801 as the sum of the first four numbers, $7+49+343+2{,}401=2{,}800$, then simply adds 1 to get 2,801, which is the same number that appears in our shortcut method. This problem is not only a matter of multiplication and addition, but it also demonstrates a far deeper understanding of the geometric series. The Rhind Papyrus is full of similar problems, dealing with quantities like loaves of bread, or bags of silver.

Ancient Egyptians had an advanced knowledge of geometric concepts. They did build the Great Pyramids after all! By the third century BCE, Egypt was the center of all mathematical progress in the Mediterranean world. This was largely due to their Great Library of Alexandria, which was home to tens of thousands of papyrus scrolls, not only about mathematics but also about an incredible diversity of topics and disciplines, from the poetry of Sappho, to the star maps of Hipparchus. The reason the library was so gargantuan was because King Ptolemy had a, shall we say, *interesting* method for acquiring new texts. Whenever a merchant came to Alexandria, any books they happened to be carrying would be taken away to be copied. After a scribe was finished copying the information word for word, the scribe would then hand the *copy* to the merchant and file away the *original* in the library!

One of the chief librarians during the Great Library's heyday was a man named Eratosthenes, who is known for being the first person to accurately calculate the circumference of the earth, which he accomplished with little more than a stick in the ground and some basic geometry. Travelers who visited Alexandria once told Eratosthenes that at noon on the summer solstice, the sun would perfectly illuminate the bottom of a well in a faraway city called Syene (now Aswan, Egypt).

This was odd, since the sun did *not* light up the bottom of wells in Alexandria, even at the same time of day on the same day of the year. Of course, there could have been multiple explanations for this. The travelers might have been lying, confused, or delirious from trekking over 1,000 kilometers. Or perhaps the sun was simply closer to the well in Syene than it was to the stick in Alexandria. Eratosthenes dismissed both explanations. He correctly assumed that the reason the sun cast a shadow in one city while being directly overhead another was due to the curvature of the earth. Moreover, he deduced that he could work out the circumference of the earth using only the angle of the sun's rays in Alexandria, paired with the distance from Alexandria to Syene. The first of these two measurements was easier to find.

To calculate the angle of the sun's rays, Eratosthenes placed a stick upright in the dirt, and then compared the length of the stick to the length of its shadow. Using this method (that modern students now call trigonometry), he found the angle to be about 1/50th of a circle (in modern terminology, that's 7.2 degrees).

What does this tell us?

If you imagine taking a cross section of the earth, with Alexandria and Syene as points on its edge, then cutting out a cake slice as thin as the space between Alexandria and Syene, the central angle of that slice would be 1/50th of a 360-degree circle (see page 187).

If the central angle between Alexandria and Syene (measured from the core of the earth) is 1/50th of a circle (or

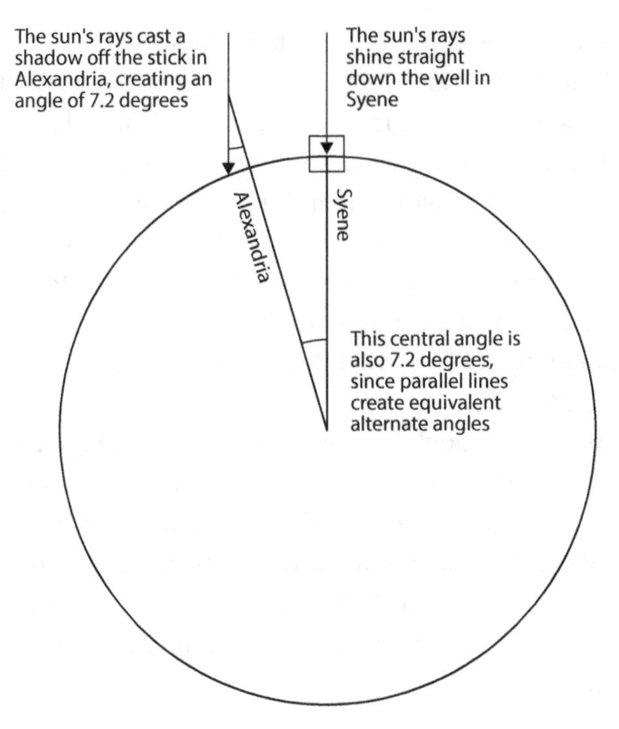

Eratosthenes's attempt at calculating the circumference of the earth.

7.2 degrees), that means that the circumference of the earth is 50 times larger than the distance between Alexandria and Syene! All that was left was for Eratosthenes to measure this distance, then multiply by 50. Unfortunately, that would prove to be a more difficult feat. Travelers would have been able to tell him how many days it took to cross the distance between the cities on a camel, but this was too imprecise. Camels might stop for a drink, go faster than other camels, or veer off from a straight path. We've all heard the truism, "If you want a job done right, do it yourself." Or, you could pay some real professionals to do it. Which is what Eratosthenes did.

He hired bematists, who were skilled in the art of walking a distance on foot with strides of exactly equal length (yes, this was an actual job description). They made the journey and reported that the distance was about 5,000 stadia, an ancient unit of measure. From there, we simply multiply by 50 to get 250,000 stadia, and this is Eratosthenes's calculated circumference of the earth! While scholars disagree on exactly how long a single stadian is, most of them think it's somewhere between 150 and 185 meters. That puts Eratosthenes's measure of the earth's circumference at somewhere between 37,500 km and 46,250 km. Considering that the circumference of the earth as determined by NASA is 40,075 km, he didn't do too badly for the year 240 BCE!

Impressive though his results may be, Eratosthenes is not even the most famous geometer to come out of Alexandria. That title goes to a man named Euclid. Euclid wrote a geometry book called *Elements*, which remains the best-selling textbook of all time, and the second most printed book in history, topped only by the Bible. To this day, every math textbook draws from Euclid as a template.

A Full Coverage Foundation

Unlike the Rhind Papyrus, which demonstrated geometric concepts using practical examples like cats and barley, the goal of *Elements* was to reduce geometry to its bare bones and prove *why* these geometric concepts worked. Euclid took apart all of geometry and reverse-engineered it. In doing so, he distilled geometry down to five fundamental axioms that he could reduce no further:

1. It is possible to join any two points with a straight line segment.
2. It is possible to extend any straight line segment indefinitely in a straight line.
3. It is possible to draw a circle with any center and any radius.

4. All right angles are equal to each other.
5. If two lines are drawn which intersect a third in such a way that the sum of the inner angles on one side is less than two right angles, then, if extended far enough, the two lines must eventually intersect each other on that side.

Euclid took these five axioms as foundational building blocks, and from there he sought to prove everything he could, like the Pythagorean theorem, the infinity of prime numbers, the formula for the area of a circle, and the volume of a cone. However, his book was not most famous for the theorems themselves but rather for his method of presenting them in a mosaic of logic that all links back to the original five axioms. When Thomas Jefferson asserted in the Declaration of Independence, "We hold these truths to be self-evident," he was referencing Euclid's axioms. Many hail Euclid as the father of geometry. I think, even more importantly, he was the father of mathematical rigor. Euclid sets the example for how people ought to approach math. The genesis of any stunning achievement is a solid starting point. Before you can build a pyramid, you must start with a solid foundation. Euclid's foundation was five axioms.

And my foundation is full coverage.

All of us use axioms in our daily lives to guide our decisions. Our personal axioms are those "default" principles that form the basis for complex decision-making. They are principles that don't require any further explanation; they're simply ingrained beliefs.

Suppose you are a parent of a young girl, and one day your daughter's teacher reveals to you that she was caught plagiarizing an essay from an online database.

Your daughter asks you, "Why is plagiarism wrong, anyway?"

You might respond by saying it's wrong because it's cheating.

"Why is cheating wrong?" she persists.

You might retort that cheating is wrong because it's lying.

"Why is lying wrong?" she asks.

At some point, you might just end the conversation with, "Because I said so."

This is where the curious kid has encountered one of her parents' ethical axioms: It is always wrong to lie.

A different parent might bestow on their children slightly different axioms, instead responding that lying is wrong only when it hurts people. Perhaps their ethical axiom is that it is always wrong to hurt people, so lies that cause no hurt are ethically neutral, and lies that prevent people from getting hurt are ethically justified. Another parent might say that the reason plagiarism is wrong is because it breaks the school's rules, and you should always show respect to authority.

It's possible to hold multiple ethical axioms, such as, "Always be loyal to your family," and "Always respect your elders," or even "Everything written in the Bible is true." You might chalk your axioms up to instinct, religion, culture, or tradition passed on to you from your parents.

I ran into one of my parents' axioms, "Always be faithful to God," when I began to question my own spiritual beliefs. A drag axiom might be, "Always entertain the audience." That's pretty self-evident and could be a solid guide when making decisions about what song to perform, what outfit to buy, or what gigs to take.

We rely on our axioms to guide us through complex decisions. In a similar way, mathematical axioms are those building blocks, accepted without question, that guide our ability to make logical arguments.

For the next two millennia, Euclid's *Elements* was the final word on geometry. Students would study it not just to learn geometry but also logic and rhetoric. Up until the twentieth century, it was considered standard reading for any educated person in England. Legend has it that Abraham Lincoln kept a copy of Euclid's *Elements* by his bed and studied it every night.

The reason Euclid's setup for *Elements* was so admired was because it was so convincing. He made it practically impossible for any reader to walk away not believing in the Pythagorean theorem. To be that compelling, his logic needed to be rock solid. Here are the two reasons why it worked:

1. It avoids circular reasoning. You can't assume the thing you want to prove. You have to justify every step using something that the reader has already accepted in a previous step.
2. The rules of engagement are clear to everyone. Euclid does a fantastic job of laying down the ground rules by giving clear definitions and five clear axioms. Euclid leaves nothing open to debate.

Despite its glory and fame, some mathematicians found one aspect of *Elements* hard to accept. Among his five axioms, the fifth one strikes most readers as particularly cumbersome. Essentially, the fifth axiom states that whenever two lines are *not* parallel to each other, they must eventually intersect, or put even more simply, parallel lines never intersect. This axiom is commonly known as the *parallel postulate*. Many readers across the centuries have found the parallel postulate too awkward and wordy to be in the same league as the other four axioms—not that they don't think it's true but rather because they feel it is not as self-evident as the first four. To reiterate the first four: (1) you can draw a line between two points, (2) you can extend a line as long as you want, (3) you can draw a circle with any center and radius, and (4) all right angles are equal to each other. The parallel postulate just doesn't feel as simple as these other four to some folks. What do you think?

The problem is that if we just ditched the parallel postulate altogether, we wouldn't be able to prove that the angles in a triangle all add up to 180 degrees, or that the Pythagorean theorem is true. So, what if, rather than an axiom, the parallel postulate were a theorem instead? Theorems are mathematical statements that are logical extensions of axioms. If we managed to prove the parallel postulate using the other four axioms, it would become superfluous as an axiom, yet it would still be useful and true as a theorem, and everything else in *Elements* would remain sound.

Mathematicians spent the next 2,000 years trying to prove that Euclid's fifth axiom was superfluous, by proving that it was somehow logically connected to the first four. One person who took great interest in this was al-Haytham, who was born in Basra in the year 965, during the Islamic Golden Age. He attempted the very first proof of the parallel postulate using a *proof by contradiction*. Essentially, he assumed that the parallel postulate was false, and then showed that this assumption led to some ridiculous contradiction. He demonstrated that if Euclid's parallel postulate was wrong, and parallel lines could indeed intersect, then it would be possible to draw a shape with only two sides, which al-Haytham dismissed as nonsense.[3]

The details behind his method didn't convince everyone, though. A century later, a Persian mathematician and poet Omar Khayyam made his own attempt. He constructed a four-sided shape with two right angles at the bottom and a pair of parallel lines. He recognized that he could prove the parallel postulate by showing that the top two angles of the shape were right angles. He assumed that this was obvious, and that the case for them being acute or obtuse angles was also nonsense.[4]

Both al-Haytham and Khayyam (and many others) failed to convince the rest of the mathematical world to dethrone the parallel postulate. It didn't stop them from trying, though!

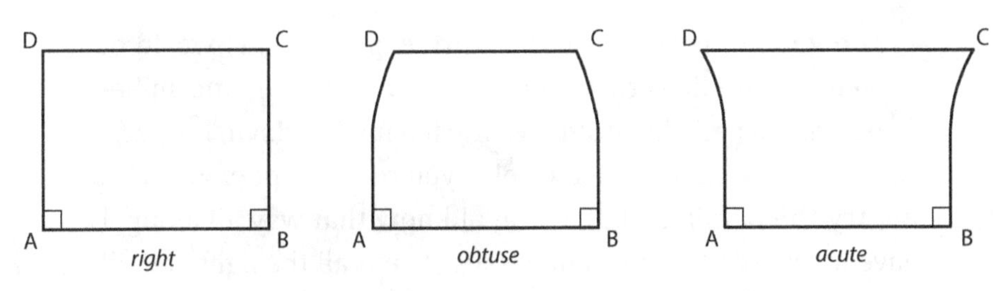

Omar Khayyam's attempt at proving that Euclid's fifth axiom was superfluous.

If there is one trait all great mathematicians share, it is a stubborn unwillingness to take knowledge at face value. What al-Haytham and Khayyam couldn't know was that their failures were ultimately anything but—in fact, they had taken the first steps toward forging a whole new branch of mathematics that did away with the problem entirely.

Rebels, Spheres, and Taxicabs

In the nineteenth century, two mathematicians, Nikolai Lobachevsky and János Bolyai, from Russia and Hungary, respectively, separately and independently made discoveries that would immortalize them as the authors of *non-Euclidean geometry*. Both of them boldly asked: What if we *didn't* accept Euclid's fifth axiom as true at all? And in true contrarian fashion, instead of removing it from the line-up of axioms completely, they exchanged it for its exact opposite. In Euclidean geometry, parallel lines never intersect because they always are a constant distance away from each other, but Lobachevsky and Bolyai made the assumption that parallel lines could curve away from each other. They postulated that the lines could even curve toward each other, and still never intersect. This new area of mathematical study became known as *hyperbolic geometry*.

Bolyai became so obsessed with exploring this world of curved straight lines that his own father, a mathematician before him, warned him against venturing further down his path: "Don't go any step further, or else you're a lost person. . . . Do not try the parallels in that way: I know that way all along. I have measured that bottomless night, and all the light and all the joy of my life went out there."[5] But Bolyai couldn't resist the thrill of the hunt. He discovered that his new axiom was mathematically sound and wrote back to his father: "from nothing I have created a new different world."[6] In the Euclidean world, it's possible to tile a two-dimensional plane by repeating the same shape, but you can only do this with squares, rectangles, triangles, or hexagons.

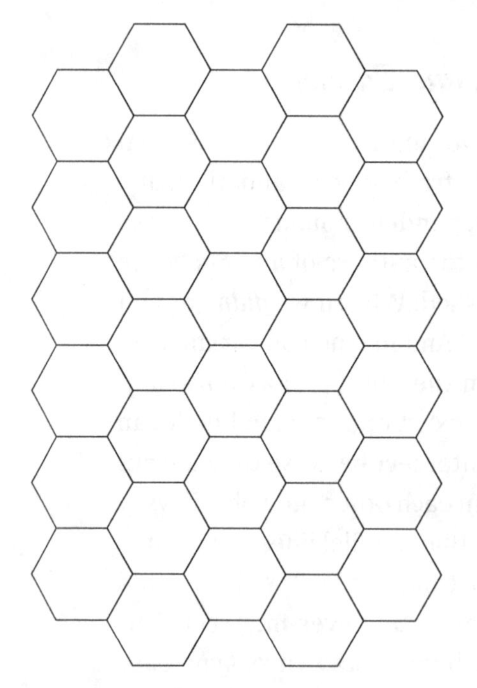

A tiling of the plane using hexagons.

A tiling of the plane using squares.

You can't tile a flat plane using a pentagon (which has 5 sides) because three pentagons jammed in a corner leaves a gap, and four pentagons creates some overlap. Tiles should fit neatly together. However, it's possible to tile the *hyperbolic* plane with pentagons.

A creative tiling of the hyperbolic plane using pentagons. To represent this tiling as a two-dimensional (Euclidean) image, we have to depict the plane as a circle and we have to show the pentagons getting progressively smaller when they are actually the same size. This type of representation is a Poincaré disk model.

We can fit three pentagons into a corner if we're willing to expand the flat surface to allow a little curvature in our plane. And don't be fooled—these pentagons are all the same size. The only reason they get smaller toward the edges in this representation is so that we can display them here on a flat page. In reality, when all the pentagons are the same size, it creates beautiful curves and ruffles, which is great if you're interested in making a non-Euclidean dress (see page 196)!

Kyne wearing her hyperbolic plane dress.
Author photo by Fabian Di Corcia.

In hyperbolic geometry, the angles in a triangle can add up to less than 180 degrees. In fact, it's possible to draw a triangle with three 0-degree angles! Every triangle in the figure on page 197 has three 0-degree angles.

I can guess what you might be thinking. Can parallel lines be curved? Don't lines have to be straight? Well, what does it even mean to be straight, anyway? I ask myself that a lot. The thing is, these lines you see in hyperbolic space *are* straight. It's the space itself that's curved!

Just as a different set of ethical axioms can produce a person with a completely different moral compass than yours, creating a different set of geometric axioms can produce a completely different geometry, with different rules, regulations, and definitions.

A Poincaré disk model of the hyperbolic plane using triangles.
Each triangle shares a corner with infinitely many triangles.

Another non-Euclidean geometry known as *spherical geometry* considers what would happen if parallel lines were allowed to intersect. You can think about spherical geometry as the geometry of walking on the surface of the earth. Picture a huge global Pride march that walks along the equator from Quito, Ecuador, to the rainforests of the Republic of Congo, at which point the parade makes a 90-degree turn from the equator and travels straight north toward the North Pole to spread some love and glitter at Santa's workshop. Then, the parade makes another 90-degree turn to travel straight down south, back to the starting point in Ecuador again. That parade will have traced a triangle on the surface of the earth with three 90-degree angles! It isn't illegal math, it's simply non-Euclidean geometry.

Another cool feature of spherical geometry is that straight lines have a maximum length. If you set off west in a straight line, you'll end up back where you started. In Euclidean geometry, axiom 1 says that lines can be infinitely long, so spherical geometry differs from Euclidean geometry in not just one but two axioms!

What else changes? In Euclidean geometry, as al-Haytham showed, it would be nonsense for a shape to have only two sides and a positive area, but this is possible in spherical geometry. The area enclosed by the Northern Hemisphere and the Western Hemisphere is a shape with two sides. Our intuitive notion of straight lines looks different on a sphere. Think about the straight line on the earth which connects China and the United States. In spherical geometry, the curved path that goes over the surface of the Pacific Ocean *is* a straight line, because it's the shortest path! Every "straight line" in spherical geometry actually traces the surface of the sphere.

Let's take for example the 16-hour flight from Toronto to Manila. The first time I took this flight to visit my family in the Philippines, I clearly remember looking at the screen on the seat in front of me to see that the outside temperature was a frigid 64 degrees below zero Celsius. (That's 83 degrees below zero Fahrenheit for the Americans in the house.) That's because, to my surprise, we were flying near the North Pole. Instead of simply flying south through the United States and over the Pacific Ocean and into Asia, our flight took us up north, into the Arctic Circle and over Russia. On a flat screen, the path looks ridiculous. That's because the shortest path between the cities on a sphere looks like a curve when we draw it on a flat plane. Straight lines in spherical geometry look like curved lines in Euclidean geometry!

But what if we're more into staycations than 16-hour flights? Surely the curvature of the earth doesn't matter as much when traveling short distances, so flat-plane, Euclidean geometry is the best model when traveling within a city, right? Not necessarily! I once went on a jaunt around New York City to film videos about the math behind some of the city's iconic landmarks, like Grand Central Terminal, Times Square, and

Wall Street's *Charging Bull* sculpture. In calculating the most efficient way to travel around the city in full drag, the straight line, Euclidean distance between landmarks was a nonissue. The real concern was the distance I had to walk in heels and the money I had to pay a cab driver. Since cabs can't drive through buildings, the only *metric* (a method of measurement) that mattered was the one Google Maps uses, which follows streets and sidewalks. That's called the *rectilinear distance*, or the taxicab distance. *Taxicab geometry*, named after the limited paths that taxicabs take in a city like Manhattan, is another non-Euclidean geometry.

This question of which geometry to abide by was actually debated in court in 2005, after a man named James Robbins was arrested for selling drugs to an undercover cop on the corner of 8th Avenue and 40th Street, which happened to be within 1,000 feet of a school. This was a germane issue because a conviction of selling drugs near a school was (and is) prosecuted much more harshly in New York State than your average small-time drug deal. The school in question was down the block and around the corner behind some buildings from the site of the shady deed. Because the prosecutors couldn't walk through walls, they weren't able to measure the exact straight-line distance to the school. Instead, they used the Pythagorean theorem. Starting from the spot where the ill-considered deal took place, they walked north along 8th Avenue for 764 feet, and then took a sharp left turn to walk west on 42nd Street for another 490 feet, until they reached the school. This created a right-angled triangle, which allowed them to calculate the hypotenuse at 907 feet. Busted! But Robbins's lawyers revealed a counterargument. The law never specified which definition of distance was to be used! According to Euclidean geometry, 907 feet is the straight-line distance between the corner and the school. But the practical distance would have measured the path along streets. By the rules of taxicab geometry, that distance measured 1,254 feet (764 + 490), which would place the drug deal more than 1,000 feet away from a school. However, the court wasn't convinced by the lawyer's argument.

It turns out that the default geometry of US law is Euclidean.

If metrics allow us to tweak the very definition of distance, then what happens to every other concept that depends on distance? Turns out something as seemingly straightforward as point *A* to point *B* can go a little queer with the practical application of a creative imagination.

What if we applied a notion of distance to something not physical at all, like words? We might say that the words *DRAG* and *DRAW* are 1 letter apart because they differ only in the fourth letter but otherwise have the same first 3 letters in common. The words *DRAW* and *DREW* would also be 1 letter apart, but *DRAG* and *DREW* are 2 letters apart.

What would the circle of words that are 2 letters away from *DRAG* look like? The circle would be a list of words like *FLAG, SHAG, DRUM, DREW,* and *FRAT*. This type of distance metric is called an *edit distance* by computer scientists, who use it to create programs like autocorrect! It can also be used to compare two DNA sequences and quantify the similarities or differences.

The paradigm shift ushered in by non-Euclidean geometers was nearly as influential as Copernicus proving that the earth revolved around the sun. In a time when everyone revered Euclid as the king of logic and geometry, nonconformist mathematicians came along and struck a blow against the status quo.

In 2023, Missouri state representative Phil Christofanelli challenged fellow Republican representative Ann Kelley on a bill she sponsored that would ban all classroom instruction relating to sexual orientation or gender identity from kindergarten all the way through high school. He asked her whether the bill would prohibit discussions of heterosexual relationships, as well, like the beloved power couple George and Martha Washington. Kelley replied, "To me that's not sexual orientation." Christofanelli countered, "So it's only really certain sexual orientations that you want prohibited from introduction in the classroom?" Kelley revealed her guiding axiom in her response: "We all have a moral compass. And my moral compass is compared with the Bible."[7]

Every law is written with the writer's personal beliefs guiding their hand. What math teaches us is that we can question those axioms and come up with entirely new systems and laws based on different axioms, and those new systems can be major improvements.

In 1915, Albert Einstein published perhaps his most famous paper on what he called general relativity. In it, he suggested that spacetime was actually curved by the gravitational pull of large bodies in space. This means that the universe itself does not operate in flat planes and straight lines but rather the more nuanced, queer, and curvaceous language of non-Euclidean geometry. Somebody needs to inform Ron DeSantis to rewrite the Florida school curriculum to incorporate the important math proving the universe is not straight!

None of this means that Euclid was wrong, by the way; he just operated by a different set of axioms. What a person believes to be morally justified depends on their personal ethical axioms and, similarly, what a mathematician believes to be true depends on their mathematical axioms.

How to Transition a Square

Mathematicians today continue to push the boundary of what we can learn with logic. One vibrant contemporary field of math related to geometry is known as *topology*. Topology is a bit more abstract than geometry. For instance, no matter what form of geometry you use, be it Euclidean or otherwise, a triangle will always have three sides. But in topology, a triangle is equivalent to a circle, a square, and a star. The traditional notion of equivalence is relaxed, so any two shapes can be topologically equivalent if they can be smoothly morphed into one another. Instead of having rigid sides, shapes are kind of like necklaces. Depending how it's stretched, a necklace can take the shape of a circle, oval, and a triangle! A cube made out of clay can be morphed into a sphere, and a coffee mug can be morphed into a donut, which can then be morphed into a skirt!

One question that gets everyone talking about topology is: How many holes does a straw have? I like to throw it out there as part of my game Math Queen Pop Quiz. Common answers are usually 0, 1, or 2. A straw with a hole in it is a faulty straw—that's the argument for zero holes. The 2-hole camp says that the straw has holes at the top and bottom, and the 1-holers say it's all the same hole. While we're at it, how many holes are in a pair of jeans, a T-shirt, or a human? How would we approach this as a topologist? Since it's theoretically possible to stretch and shrink a straw into a donut, it has the same number of holes that a donut has: 1!

You might be wondering right now, as every student eventually does, what on earth this wild mathematical concept has to do with real life. The answer is everything, darling! Techniques developed in topology are used to understand the shapes of DNA, computer networks, and even the universe itself.

But you should know by now that all the best math was formulated to satisfy curiosity, and more importantly, for fun! Mathematicians and drag queens all use make-believe to reveal hidden truths about reality. Drag queens use costumes, smoke, and mirrors to convey truths about the human condition, and mathematicians use abstraction and imagination to convey truths about nature. Bending and stretching the mathematical imagination is crucial to its development.

Geometry has always straddled the line between imagination and reality. What we can theorize in our minds about hypothetical lines and shapes, we can apply to reality, whether by folding paper, constructing pyramids, measuring the earth, or navigating great distances. We can start with small building blocks to create complex logical arguments and broaden our vision so vastly that we can study the geometry of the universe itself.

Just as non-Euclidean geometries challenge the traditional notion of what geometry ought to look like, queer identities challenge our notions of how humans ought to express themselves. One of the reasons I love geometry is because it queers ideas we have taken for granted as fundamental truths: what a parallel line really looks like, what a circle looks like, and what it means to be straight or curved.

Most students leave geometry class remembering little more than how to mimic a phrase like $a^2 + b^2 = c^2$, but triangles, parallel lines, and the Pythagorean theorem are only the grammar and punctuation rules of geometry. Sure, they're important, but you wouldn't expect someone to love reading after taking a class on grammar. In the fashion world, you have to learn about pattern-making and measurements before you can work your way up to a Bob Mackie gown. In music, you have to practice scales before you can sing like Donna Summer or play the flute like Lizzo. But in math class, students think that the tedious part is all there is, because all the brightest, most exciting stuff is hidden deep in postgraduate degrees! Students aren't learning about the fantastic side of math where we can create our own universes, ones with totally different axioms, where triangles can have three 0-degree angles, parallel lines can curve toward each other, and the Pythagorean theorem is all wrong.

The most important takeaway is that geometry is fundamental to logical thinking, but it's still boundary-breaking, philosophically exhilarating, and queer as hell. When we question tradition and question our axioms, we may find that the framework we've used to view the world wasn't necessarily the right one, and it can be swapped with other, equally valid frameworks that allow new ways of viewing reality.

Our world would not be where it is today without those who challenged tradition and stretched the norms. What we got in return is a universe that is all the more colorful, vivid, and powerful. Our families, schools, and books should teach young people that we must continue to challenge traditions, stretch norms, and open our minds. It will make them better mathematicians and better people.

CHAPTER 9

Mathematical Realness

The concept of "realness" can mean different things depending on whether you're a philosopher, a mathematician, a drag queen, or none of the above. Are numbers real? Does math exist in objective reality, or is it merely a figment of our collective imagination?

It's a matter of perspective. Numbers may not be tangible, but they are a real human construct, and imagining a world without them is nearly impossible.

For drag artists, *realness* is a specific concept: it is the measure of one's ability to blend in with the "real" world, i.e., the world of straight, cisgender, white people. This notion of judging realness has its roots in the early days of queer, Black ballroom culture in the United States. A highlight of the first drag balls incorporated a popular dance of the period, called the prize walk, or cake walk, which originated among enslaved people on plantations, and may have begun as a mockery of the formal European-inspired dances of plantation owners. Attendees at the drag balls would dance or "walk," in their dazzling gowns, with the most graceful and skillful winning a cake as a prize. Although ballroom persisted as a subculture in the United States for the next century, especially in New York City, it was the queens of the vibrant scene of 1970s Harlem who began to codify the elements of drag balls in a new way.

In the late 1960s, legendary queen Crystal LaBeija founded the first modern ballroom house in Harlem out of exhaustion and frustration with the racism she faced in the traditional drag pageant circuit. The first drag documentary, *The Queen*, captures LaBeija calling out racist corruption in the Miss All-America Camp Beauty Contest, when an inexperienced and unpolished white queen wins the crown. She is visibly exasperated as she insists, "Miss Thing, I don't say she's not beautiful, but she wasn't looking beautiful tonight!"[1]

Sick of the way Black queens were mistreated, overlooked, and expected to use complexion-lightening makeup in order to be serious contenders, LaBeija started her own pageant: the House of LaBeija Ball. As the first-ever mother of a house, she started a drag revolution. Mixing elements of pageantry and ballroom tradition with the unique Harlem queer culture of the time, the new balls centered Black and Latina queens. The competitive spirit of the pageants and prize walks remained an important aspect of the balls, and when a new dance called voguing emerged, which took inspiration from the acrobatic poses of supermodels in *Vogue* magazine, the walks evolved to include vogue battles.

Along with Vogue Performance, three more competition categories—Fashion, Face, and Runway—became key in this emerging ball scene, as did the artistry known as realness. Realness is assessed within multiple contexts. In the category of *femme queen realness*, participants are judged on their ability to pass as "real" women, that is, straight, cisgender women. *Butch queen realness* tests one's ability to blend in with straight, cisgender men by adopting their appearance, mannerisms, and demeanor. The category of *executive realness* sees participants pose as if they had just come off of the streets of corporate America, flaunting a three-piece suit, briefcase and all. If this seems like a less than obvious category, consider that for most of the participants, who were predominantly Black trans women and gay men, being an actual executive was out of reach, requiring expensive education and familial connections. Ballroom gave them a place where they could be triumphant and free, and pretend for one night that they could live among "real" men and women and be given the opportunity to have "real" jobs. As the famed queen Dorian Corey described it, "If you can pass the untrained eye or even the trained eye and not give away the fact that you're gay, that's when it's real. . . . The idea of realness is to look as much as possible like your straight counterpart."[2]

To drag queens like me, realness is a matter of performance. I impersonate a woman on stage and on camera, but live the majority of my life as a boy. Unlike professional actors who take on many different roles throughout their careers, my drag character is a singular extension of myself, like an alternate version of my personality. It's hard to decide which version is my "real" self. To express my true, uninhibited style would mean for me to show up with long hair and colorful makeup in a sequin dress, as well as a constricting corset, heavy earrings, and many layers of duct tape and pantyhose. When I take off all the layers of my costumes, I'm left with my other "real" self, the version of myself that I take to bed, to doctor appointments, and to dinner with my family. I haven't decided yet which version of Kyne I should be buried as. It'll probably be whichever one I'm in when I die. . . . On second thought, maybe I'll go for cremation.

In some sense, priests dress in drag every time they put on their robes. Doctors get in drag by putting on their scrubs. Customer service reps perform in drag every time they talk to consumers. Soldiers, police officers, judges, pilots, sex workers, and bankers all wear costumes and disguises, playing performative roles that function as extensions of their personalities. As RuPaul put it, "We're all born naked and the rest is drag."

So what does it really mean to be real? In the context of ballroom, realness means white picket fences, nuclear families, and a white-collar job. Realness refers to the people and the lifestyles that the dominant culture endorses as legitimate and worthy of dignity and protection. If you don't belong to one of the identities that the legal system, politicians, taste-makers, and average people deem real, then you have to speak in coded language and move in the shadows.

When Jennie Livingston released her documentary *Paris Is Burning* in 1990, ballroom entered the mainstream. The documentary follows the lives of iconic members of the Harlem ballroom scene, including Dorian Corey, Crystal LaBeija, Pepper LaBeija, Angie Xtravaganza, Willi Ninja, and Paris Dupree, whose annual ball inspired the film's title. That same year, Madonna released her song "Vogue," which climbed to number one on the charts and introduced the world to the art of voguing. The ballroom scene was put in a huge, bright spotlight, unlike ever before. Madonna, at the peak of her career, hired ballroom dancers to choreograph her music video and to dance on her world tour. Ballroom culture went on to inspire fashion, dance, and drag styles, and has made its way into everyday language, with phrases like "throwing shade," "fierce," or "werk!" A huge chunk of the lingo that's currently associated with LGBTQ people or with Gen-Z actually originated in Black American ballroom.

All the newfound attention and interest in ballroom in the 1990s happened at the height of the AIDS epidemic. The disease claimed the lives of many members of the LGBTQ community, including many of the most talented and prominent ballroom elders. Gay men and trans women of color have been the most disproportionately affected by HIV/AIDS, and if queer people weren't already othered enough, the AIDS epidemic added a new layer of stigma. Many members of the public viewed it as a "gay disease" and/or God's punishment for living as a sexual deviant.

The government failed to properly educate the public about risks and prevention, in no small part because they did not want to explicitly address the queer community. As a result, many HIV-positive people swept their diagnosis under the rug instead of seeking proper treatment and support. The queer elders we still have are living reminders of how far we've come in the span of a single generation, and how much more work still needs to be done to achieve visibility and equity. The ballroom community needs not only to be recognized for their cool dances and popular slang, but for the reality of their experiences, past and present. Their existence is not predicated on entertaining viewers; they are humans who deserve dignity, reparations, and support. They are, in every sense of the word, real.

Over a hundred years after the advent of drag balls, queer people as a whole still struggle with being seen as real. We're told that real boys don't wear makeup and dresses and that real girls act ladylike. Anyone who defies the gender binary is cast as confused or mentally unstable, and is told to get back in line with reality. Those of us who dare to identify as a nonbinary gender or use alternative pronouns like "ze/zim/zirs" are accused of inventing fake genders and fake science. There are plenty of people even within the gay community who feel that nonbinary genders are not real, nor are alternative sexual orientations like pansexual or demisexual.

Who decides which identities are real and which aren't? Who dictates what a "real" boy or girl should look like? Is realness a matter of physical appearance, or is it a feeling in your bones?

Realness, as it's seen from the lens of the ballroom category, is the state of being embraced and recognized as legitimate by broader society. In the past, people thought that the only real or legitimate trans people were those who made permanent alterations to transform their body, and they were called transsexuals. Today the term "transsexual" is largely seen as outdated, although I still know some trans people who identify with it. The more commonly accepted name for trans people is transgender. Not all transgender people wish to pursue medical or surgical transitions. Others have suggested that the only real trans people are those who are diagnosed with gender dysphoria by a medical professional. There are a couple of problems with this view, the first being that some trans people don't experience gender dysphoria but still identify with the community. Second, there are financial and legal obstacles to seeking a diagnosis, especially with politicians making it harder for people to seek gender-affirming medical care, or even learn about the vocabulary to describe what they might be feeling. Despite all that, trans people are, and will always remain, valid and real.

In many progressive societies, your "real" gender is a matter of legal paperwork and checking the right boxes. This notion seems to make some people uncomfortable. If realness is merely a matter of self-identification, does that mean the notorious white conservative Piers Morgan can identify as a Black lesbian, as he claimed on his television show? Can conservative trolls begin identifying as attack helicopters? Can we also start saying that $2+2$ is equal to 5, or that $1/0$ is equal to π? If we are inclusive of every identity that anyone speaks into existence, then how do we know which identities really exist?

Realness isn't just a matter discussed in the world of ballroom, but in all of the arts and sciences. Philosophers have spent thousands of years discussing the nature of reality and truth.

The school of thought called materialism states that the only "real" things are those that we can touch, hear, taste, smell, and see. That would mean that flowers, mountains, planets, and animals all exist, but abstract concepts like genders, thoughts, and mathematical equations don't. Where, then, do numbers and shapes fit in? If flowers exist, and two flowers exist, then does the number 2 exist? How about negative numbers, irrational numbers, or imaginary numbers? How about infinity? Where do we draw the line between physical reality and abstraction? Shapes also straddle this line; we can clearly observe circles and triangles in nature, but zooming in on the edges of any shape in nature will reveal bumps, holes, and imperfections. Math works with utopian models of perfectly smooth circles, and perfectly straight lines that stand up to infinite zoom. One might argue that these mathematical objects are inventions of the human mind and are therefore not real. But if math indeed is a human invention, then it's an invention that was developed by various societies around the world independently of one another. The number 0, for instance, was invented in China, India, Mesopotamia, and Mesoamerica, all separately. Perhaps zero was not *invented* but instead *discovered*.

If math is an imaginary creation, it is one that comes with preestablished rules. Once the numbers and basic structures are put in place, consequences follow and can't simply be changed according to our whims. That's why we can't start saying that $2+2=5$ and that $1/0=\pi$. There is an order in math that we accept as inherently definitive and beyond our control, yet at the same time lives inside our brains alongside all our other jumbled, and sometimes contradictory, thoughts.

The Greek philosopher Plato believed that numbers existed as objective truths in a world of forms, a place beyond heaven that humans had no control over. According to Plato, mathematical truths were eternal and independent of humans. But not all philosophers agreed with him. The wide dissemination of non-Euclidean geometry marked a watershed moment not only for mathematics but also in the philosophy of mathematics. While we once took Euclidean geometry for granted as a basic truth handed down from God herself, the existence and validity of non-Euclidean geometry flipped reality upside down. The idea that truth depends on axioms defies the notion of math being dependent on objective truth. The Pythagorean theorem, for instance, is true only in the sense of Euclid's five axioms, which by definition are unprovable statements that one must either accept or reject. The theorem is only true in the same sense that "Spiderman is Tom Holland" is true. The identity of the actor behind Spiderman depends on which movies you watch; the truth value of the Pythagorean theorem depends on which geometric axioms you use.

The statement "I am a man" is only true depending on our definition of "man." Definitions of words are social constructs that may vary somewhat from person to person or from one culture to another, but have a generally set meaning that is only meaningful through the lens of an agreed-upon, objective reality. Definitions can change, though, such as the word "gay," which once meant "jovial," then morphed into a euphemism for homosexual men.

When I was young, my classmates would use "gay" as a general derogatory term for anything uncool, corny, and lousy. The word "queer" has also gone through significant changes, originally meaning "strange" or "peculiar," then becoming a pejorative for gay men. Many young people have reclaimed "queer" as an umbrella term for everyone who lives a nonheteronormative lifestyle. A sentence like, "I went to a gay party," can have multiple meanings, depending on the definition of "gay." Sometimes shifting definitions can unsettle people. It's human nature to value the security that comes from certainty. That's understandable, and okay! Except when clinging to what we've been so sure about stifles our ability to entertain innovation and accept greater wisdom.

The non-Euclidean revelation made a lot of mathematicians very uncomfortable because they felt that without a strong logical foundation of objective reality, suddenly nobody was sure what mathematicians were really studying. If any statement could be true or false depending on the set of axioms and definitions used, then how could anyone be sure that their work had any real meaning or substance to it? How can we be so sure that $2+2=4$?

Building Math from Scratch

In the nineteenth century, mathematicians began a quest to start from scratch and reverse-engineer all of mathematics just as Euclid had done to geometry. Some mathematicians thought that axioms could be used to describe everything in mathematics. This effort gave birth to a new branch of math known as formal logic. Up until now I've been using the word "logic" in the context of its everyday meaning, which has to do with commonsense reasoning and rhetoric. Formal, mathematical logic is the study of all the rules, axioms, and language used to describe math itself. Meta, I know!

Consider the following three sentences as an example:

All drag queens wear blue eyeshadow.
Anna is a drag queen.
Therefore, Anna wears blue eyeshadow.

This kind of argument is known as a *syllogism*. The first two sentences are premises, and the third one is a conclusion. Now assume you are told that Carmen is also a drag queen. Can you conclude that Carmen wears blue eyeshadow? Yes, because of the premise "all drag queens wear blue eyeshadow."

What if I told you that Lucy is someone who wears blue eyeshadow. Can you conclude that Lucy is a drag queen? Not necessarily. Just because all drag queens wear blue eyeshadow doesn't mean that all blue eyeshadow wearers are drag queens. Kind of like how all thumbs are fingers but not all fingers are thumbs. What if you are told that Victoria does *not* wear blue eyeshadow? Can you conclude anything from that? The answer is yes. The fact that Victoria doesn't wear blue eyeshadow is enough to disqualify her from being a drag queen. At least according to our rudimentary syllogism.

These are the kinds of structures that logicians study, along with the *inference rules* that tell you exactly how you can conclude information based on given premises. Some mathematicians thought that all of mathematics could be distilled down into logical statements, and this school of thought is called *logicism*. (Note that not all logicians are logicists.) Logicists believed that all of mathematics could be boiled down to logical rules and axioms. That includes theorems about triangles, prime numbers, sets, probability, and infinity. Logicists hoped that formal logic could provide the solid foundation to mathematics (that they wanted so badly to rediscover) and answer the questions they had been struggling to solve.

One of these logicists was Gottlob Frege. To describe all of mathematics using logic, Frege first had to define what a number was. He tried doing this without any intuition, common sense, or experience—only logic. He did so by presenting a theory of what he called concepts and extensions. Concepts were about as vague as they sound. A spoon could be a concept, and the corresponding extension would be all spoons. If the concept is "things made out of silver," then the corresponding extension would be the set of all things made out of silver, including spoons, jewelry, electronics, etc. Frege thus defined numbers as extensions of concepts. Start with an empty concept, and then add in it two apples, two rocks, two queens, your left and right earring, your left and right eyelashes, and your brother and your sister—in effect, every possible collection of two things. The corresponding extension of this weird concept is the thing that all these object pairs have in common: the number two. And that's how Frege constructed numbers out of thin air.

Frege was on to something. His ideas were well ahead of his time; many of his contemporaries could not even understand his work, but it has been a huge influence on the modern language of logic. Only one of his contemporaries could go toe-to-toe with him, a British logicist named Bertrand Russell. One day, Russell wrote Frege a letter pointing out an inconsistency he found in the theory of concepts and extensions, which raised a major problem.

To understand why, we need to think back to set theory. Russell pointed out that Frege's theory relied on every possible concept having its own extension. These extensions were sets. Consider the concept of "people." The corresponding extension would be the set of all people. That set *doesn't contain itself* because a set is not a person. That's fine, but now consider the concept of "sets" itself. The corresponding extension is the set of all sets. The set of all sets is itself a set, so it *must be a member of itself*. So, we see that some sets contain themselves, and others don't. Okay, still no problem—except that Russell pointed out the concept of "sets which do not contain themselves as a member." The corresponding extension to this concept is then the set of all such sets which do not contain themselves. Russell's seemingly simple question followed: Does this set contain itself?

If the set does *not* contain itself, then it *must* contain itself because it's the set of all sets that don't contain themselves. On the other hand, if the set *does* contain itself, then by its own rule, it must *not* contain itself. Russell's paradox points out the inherent problem of self-referencing.

With one letter, Russell rendered Frege's entire life's work meaningless. Shortly after receiving Russell's letter in 1903, Frege's wife, Margaret, died. These two events together seem to have broken Frege; he spent the remaining two decades of his life in a deep depression, dedicating his final years to detailing his anti-Semitic attitudes in his diary.

Even though Frege gave up, other mathematicians carried on the torch of logicism. But some started to have second thoughts, doubting that logic was enough to underpin all of mathematics. An alternative philosophy called *intuitionism* held instead that math was a combination of both logic and intuition. According to the intuitionists, logic was used to organize mathematical sentences in a coherent way, but the basic ideas spring from our own minds.

The intuitionist school was founded by Dutch mathematician L. E. J. Brouwer, who believed that mathematics was nothing more than a creation of the mind. Math only "exists" in the real world as much as fairy tales exist; when no one is around to think about them, they cease to exist. Importantly, the intuitionists believed there was no such thing as an objective mathematical truth. They held that we only accept a statement as true when it personally convinces us, which means for intuitionists, truth is fundamentally subjective.

While the intuitionists believe that math is only a creation of the mind, another school of thought called *formalism* posits that math is about nothing at all. Formalists believe that math is merely a large game of symbols that we manipulate on paper, like tic-tac-toe or chess. You start with your axioms, which dictate which pieces can move where and which moves are legal, and then manipulate the pieces according to the rules and see what happens. If you change your axioms, you will be playing an entirely different game. Under formalism, there is no objective mathematical truth without predetermined game rules. Formalists didn't believe in the existence of numbers either. Numbers were simply symbols with an agreed-upon meaning. When you write $1+2=3$ or $10^3=1{,}000$, you aren't discovering a new truth about the universe, you're simply moving around symbols, which is exactly what math feels like to a lot of us.

The formalists had a lot in common with logicists in that they both wanted to construct a full, logical framework that described all of mathematics, with reasonable axioms, inference rules, and a formal language. But for David Hilbert, one of the best-known formalists, the goal of establishing axioms wasn't to turn intuitive sentences into logical formulas. Rather, it was to create a powerful language free of contradictions. Hilbert was a fixture at the University of Göttingen, which boasted Europe's most dominant academic math program of the era (and where, as you might remember, he spent some years covering for Emmy Noether). His aim was to discover an optimal set of axioms and rules to frame mathematics as a perfect game that would fulfill his three requirements: completeness, consistency, and decidability.

By *completeness*, Hilbert meant that he wanted every question to be deducible from the axioms. That is, every mathematical proof could be expressed as a finite string of symbols, and every sentence could be proved as either true or false. If there were sentences with truth values that were impossible to deduce from the axioms, then the language was incomplete, like a chess board with inaccessible squares. On the other hand, if the language could prove that a sentence was both true and false at the same time, then the language was inconsistent, and some game pieces would need to be removed or modified to avoid contradictions. Hilbert knew that for the language to have any meaning or usefulness, it would need *consistency*. Finally, Hilbert wanted his language to be decidable. He wondered if there was a systematic way to find every proof. Some proofs, like the proof that π was irrational, took centuries to find. Fermat's *last theorem*, another vexing question in math, wasn't resolved until 1995, a jaw-dropping 358 years after it was first articulated. If a formal language had *decidability*, it meant that no proofs were impossible to find—an algorithm existed for finding every one of them.

Unfortunately, Hilbert's dream was doomed, just like Frege's. A 24-year-old mathematical luminary named Kurt Gödel proved that any formal system of math worth using will always be either incomplete or inconsistent, by pointing out a paradox of self-reference just like Russell had. Gödel showed that Hilbert's formal system failed when it handled mathematical statements about the system itself, such as the sentence, "This sentence is unprovable." If this statement is true, then we have a true statement which is unprovable. If it is false, then it must be provable—but if it's provable, then it must be true! Either way, Gödel proved the existence of true, unprovable statements in math, meaning that Hilbert's system, and any other formal system, would be inherently incomplete.

Appropriately, Gödel called this discovery his *incompleteness theorem*. Gödel's incompleteness theorem didn't apply to *all* mathematical systems but, as I mentioned, it does apply to those systems that are worth using, meaning systems that are at the very least capable of proving some elementary arithmetic like addition and multiplication. A system that could pass the test for being both complete and consistent does exist; it would just be too weak to even describe the natural numbers. That wasn't much consolation to Hilbert.

If we are willing to accept a little bit of incompleteness, could we at least perhaps develop a language that is consistent and decidable? Not if Gödel has anything to say about it. He wasn't done wreaking his havoc on the math world yet. While his first incompleteness theorem demonstrated that some true statements will be unprovable, Gödel proved in a second incompleteness theorem that a system cannot prove its own consistency. French mathematician André Weil put it succinctly when he said, "God exists since mathematics is consistent, and the Devil exists since we cannot prove it."[3]

Gödel's incompleteness theorems were a shocking plot twist that turned foundations of mathematics into a whole new field of study. More so than ever before, mathematicians were asking themselves: What is the real nature of truth? Is truth independent of human thinking, or is it embedded within it? Is truth itself even real? If as Gödel posits, there are true statements whose truth cannot be verified by a proof, then are they really true?

The state of mathematics as Gödel left it was this: out of all the mathematical statements we can articulate, some statements may be neither true nor false. Out of all those sentences which *are* true, some true statements are unprovable, so their truth value can't be verified by us. But leaving the unprovable aside for a moment, how about a sentence whose truth value *is* provable? Is there a systematic way to find that proof? This brings us back to Hilbert's problem of decidability, or as he put it in his native language German, *Entscheidungsproblem*, or the *decision problem*. Consider the following example:

Riddle: Two trains are traveling west toward a city on parallel tracks, and start from the same place at the same time. Train A is traveling at an average speed of 100 km/h, and train B travels at an average speed of 80 km/h because it makes more stops. If train B arrives at the final stop 1 hour after train A arrived, how far from the final stop was the starting position of both trains?

Solution: If you don't know where to start, you could always try *the guess-and-check method*. Let's guess an answer of 500 km and then see where that takes us. Traveling 500 km would take train A 5 hours and train B 6.25 hours. That's a difference of 1.25 hours, which is a little bit off, since we want their arrivals to be spaced apart by only 1 hour. Let's aim our guess a little bit lower at 400 km. That would take train A a total time of 4 hours and train B a time of 5 hours which puts them an hour apart! So 400 km is the right answer.

If you happen to be more experienced with questions like this, you might know an algebraic method for solving it which uses equations and variables:

(1) Let X be the total distance (in km) traveled (this is the answer we're looking for).
(2) Let T_A be the length of time (in hours) it took for train A to arrive at the final stop.
(3) Let T_B be the length of time it took for train B to arrive at the final stop.
(4) Then $X/100 = T_A$ and $X/80 = T_B$, because distance/speed = time.
(5) Also, $T_A = T_B - 1$, because train A arrived 1 hour earlier than train B.

Substituting $T_A = T_B - 1$, we have:

$$\frac{X}{100} = T_A = T_B - 1 \text{ and } \frac{X}{80} = T_B$$
$$\frac{X}{100} + 1 = T_B \text{ and } \frac{X}{80} = T_B$$
$$\frac{X}{100} + 1 = \frac{X}{80}$$
$$\frac{4X}{5} + 80 = X$$
$$80 = \frac{X}{5}$$
$$400 = X$$

This step-by-step procedure works for many kinds of problems, but not all of them. As you might have gathered by now, lots of mathematical questions don't have a clear step-by-step solution. Contrary to what many of us are taught in school, math is not a set of ironclad rules that always result in a single correct answer to any problem.

The class of complex problems that inspired Hilbert are called Diophantine equations, named after Diophantus of Alexandria. Diophantus was interested in finding whole-number solutions to multivariable equations, like $a^2 + b^2 = c^2$. So even though substituting $(\sqrt{2}, \sqrt{3}, \sqrt{5})$ into this equation for (a, b, c) would technically solve things, it wasn't a whole number solution that Diophantus was looking for. During all the centuries mathematicians have studied Diophantine equations, there was never a single step-by-step procedure for finding these integer solutions. Many of them were just found by the guess-and-check method. Hilbert wanted to know, does a step-by-step procedure even exist for Diophantine equations? More generally, does a step-by-step procedure exist for answering *every* question we can ask? Hilbert was hoping that every question had a systematic answer that didn't require any intuition, but instead relied on algorithms.

An algorithm is a step-by-step procedure that carries out a task. It can look like a list of instructions, a flowchart that handles different cases and contingencies, a recipe, or even just a routine. We used an algorithm to solve the traveling train word problem. You might have an algorithm for doing your makeup, which looks like a list of steps: apply eyeshadow; powder your face; if it's sunny, apply SPF first; if I'm going somewhere fancy, use a darker eyeshadow. Most people at this point are familiar with social media algorithms, which customize your experience by recommending content you personally are most likely to engage with based on your search history, the accounts you follow, your location and demographics, and perhaps even your activity on other websites. Hilbert belonged to a time before the internet though. In fact, it was Hilbert's decision problem that motivated the development of the modern computer in the first place! But that massive mathematical leap forward required a significant shift of focus and venue.

After World War II, the dominance of straight, white, primarily German, mathematical innovation seems to have rapidly declined. This is no accident. It is widely agreed that the Nazi expulsion of Jewish intellectual leaders had devastating impacts in many tragic ways, including the loss and displacement of mathematical brilliance. A story about our favorite formalist, David Hilbert, goes that he was once seated next to Hitler's education minister at a university dinner, and when the minister asked him, "How is mathematics at Göttingen, now that it is free from the Jewish influence?" Hilbert replied, "There is no mathematics in Göttingen anymore." Perhaps it is all the more appropriate, then, that at this critical moment the incomparable Alan Turing enters, stage left.

Halt! Who Goes There?

Computer science was revolutionized by the British mathematician Alan Turing. He's best known for breaking German ciphers during World War II, but his mathematical gifts were obvious well before his time in military intelligence. In fact, he answered Hilbert's decision problem when he was only 24 years old. He started by formalizing the idea of an algorithm by defining it to mean "anything we can teach a computer to do." Back then, computers looked nothing like the modern, omniscient robots we call computers today. In Turing's time, the word "computer" didn't refer to a metal box filled with wires and microchips—it was a job title. You had bakers, plumbers, astronomers, managers, and computers. Electronic computers weren't commercially available yet, so if you had a lot of math to calculate, you would hire a person to do it. Computers were usually women who were assigned long, tedious, step-by-step procedures—perfect for Turing's thought experiment.

Turing imagined a hypothetical, automated computing machine paired with a long piece of tape of any length (even infinite) printed with a sequence of squares, each bearing one symbol. The machine would read the tape one symbol at a time, while following a table of instructions, which included different "states" the machine could be in. The different states dealt with different cases and contingencies, just as the hand that does your makeup may be in a blending state, correcting state, or halting state once you're finished. The machine was only allowed to either: erase the symbol it was reading, write in a new symbol, move left or right onto the next symbol, or enter a new state. When the computation finished, the machine would stop, thus entering the halting state.

For example, you could feed the Turing machine a long string of 1s and 0s and ask it to count the number of 1s, or whether the number of 0s is a multiple of 3. If you got creative, every step-by-step procedure could be coded into a string of symbols and fed into a Turing machine. Remember, every electronic image, video, or message you've ever sent or received over a computer was just a long string of 1s and 0s. As such, every list of instructions that you could give to a human computer, you could theoretically convert into a string of 1s and 0s to feed to a Turing machine. That list of instructions is what we now call a computer program. In principle, a Turing machine is capable of internet searches, artificial intelligence, social networking, photo editing, and anything computers can do today. Our computers may be equipped with sophisticated materials and huge memories, but the fundamental theory of programming is the same as what Turing laid out almost a hundred years ago. The Turing machine was the blueprint for the modern computer.

Ideally, a Turing machine halts when it accomplishes its task. But some programs may cause it to loop endlessly, like the following:

If currently in state 1 and the tape reads a 1, then erase it, write a 2, and change to state 2.
If currently in state 2 and the tape reads a 2, then erase it, write a 1, and change to state 1.

If this Turing machine ever reaches state 1 and reads a tape that has a 1 on it, then it will get stuck in a loop of changing the 1 into a 2, and then changing the 2 into a 1, and so on and so on forever, without halting. A clever programmer may be able to read the instructions and spot a flaw that may cause the machine to enter an infinite loop, but a crucial question is, would it be possible to code a Turing machine to spot these flaws? In other words, could we create a Turing machine whose job is to determine whether another Turing machine will halt or not? If this were possible, it would be awfully useful in solving other major mathematical problems, including one of the biggest unsolved questions in math, the *Collatz conjecture.*

Named after its originator, Lothar Collatz, it's also known as the $3x+1$ problem, or the Kakutani conjecture, in reference to twentieth-century Japanese mathematician Shizuo Kakutani's work on it. The Collatz conjecture considers a game in which you start with a positive integer and apply the same two mathematical calculations on it over and over:

1. If the number is even, divide it by two.
2. If the number is odd, multiply it by three then add one.

Let's play this game starting with the number 1. It's odd, so we multiply by three then add one, to get an answer of 4. Apply the same rules again, now with 4. Since 4 is even, divide by 2, and the answer we get is 2. The number 2 is even, so we divide by 2 again to get back to 1, right where we started.

If we start with the number 12, the chain goes like this:

$$12 \rightarrow 6 \rightarrow 3 \rightarrow 10 \rightarrow 5 \rightarrow 16 \rightarrow 8 \rightarrow 4 \rightarrow 2 \rightarrow 1$$

The reason we stop at 1 is because 1 leads to a loop of $1 \rightarrow 4 \rightarrow 2 \rightarrow 1$ over and over. These sequences of numbers are sometimes referred to as "wondrous numbers."

Try it out for yourself! You can start with *any* positive whole number. Every positive whole number we've ever tried always leads back to 1! But we don't know whether *all* positive whole numbers will lead back to 1, which is why it's called a conjecture: it hasn't been proven yet. If you can find a number which *doesn't* lead back to 1, then you've proven the Collatz conjecture false. You may not be treated to a lavish feast like a Renaissance math dueler, but notorious mathematician and couch-surfer Paul Erdős did leave a $500 bounty on the table for you when he died, and you'll be sure to enjoy adoring acclaim from mathies everywhere. And if you can do it by age 23, you'll be one-upping both Gödel *and* Turing!

How might we go about solving this question using a Turing machine? Suppose we make a Turing machine and name it Cole, and we feed it an infinitely long tape which encodes all the positive whole numbers, starting from 1, and then ask it to compute whether each number leads back to 1 by following the rules. If Cole ever finds a number that generates its own loop just like $1 \rightarrow 4 \rightarrow 2 \rightarrow 1$ but *doesn't* run into the number 1, then you can consider the problem solved, and have Cole halt. Go collect your prize!

Otherwise, you could be waiting until the end of infinity.

For Alan Turing, waiting an infinitely long time was not an acceptable option. He wasn't interested in hiring any immortal monkeys. Turing was only interested in algorithms that could finish within a finite length of time. Likewise, the only information we're really interested in is whether Cole will *ever* halt, and we don't want to wait forever to find out.

So suppose we had another Turing machine to keep an eye on Cole and determine whether it would halt or not. Suppose we name this new machine Haltina, and Haltina can read the code of any other Turing machine and determine whether or not it would halt. If Haltina existed, then we could feed Cole's information into Haltina and find an answer to the famous Collatz conjecture. If Haltina finds that no, Cole does not halt, that means the Collatz conjecture is true, and all numbers eventually lead back to 1. If Haltina finds that yes, Cole does halt, then that means the Collatz conjecture is false; Cole has found a counterexample to prove it so. Haltina could potentially answer all sorts of open questions in math, like the exact probability of getting an unwinnable shuffle in Solitaire, or whether there are any odd perfect numbers (a perfect number is one which is the sum of all its factors excluding itself, like 6 which is equal to $1+2+3$, or 28 which is equal to $1+2+4+7+14$). All these questions can be encoded into Turing machines.

If a machine like Haltina really existed, then it could answer Hilbert's decision problem, since it could theoretically answer every mathematical question with a yes or no. But Turing demonstrated that Haltina is an impossible machine. Haltina is an algorithm that cannot exist.

To demonstrate the impossibility of Haltina, Turing supposes that we create a new, larger Turing machine called Anti-Haltina. Anti-Haltina has all the classic features of Haltina, but with two new twists. The way it works is that you feed Anti-Haltina the code of another Turing machine, and if Haltina reports that yes, that code will halt, then Anti-Haltina's new instruction is to loop forever. If, on the other hand, Haltina finds that the code in the third machine will loop forever, then Anti-Haltina's new instruction is to halt. In essence, Anti-Haltina will do the opposite of whatever Haltina predicts its input will do. So what would happen in this scenario if we fed Anti-Haltina into itself? If this code eventually halts, then Haltina will report that outcome, but the new instruction is then for Anti-Haltina to do the opposite—loop forever.

If Anti-Haltina loops forever, then Haltina will say, "No, this doesn't halt," in which case the next instruction is for Anti-Haltina to halt. This is a contradiction! A Turing machine can't run forever and halt at the same time. It's the same circle of self-referential logic that we ran into with Russell and Gödel. This Turing machine can't exist. Turing proved that there isn't necessarily a Turing machine that can solve every mathematical problem, thus proving that there are some questions that algorithms can't solve, even if we use infinitely long pieces of tape or wait infinite periods of time. Some problems can't be solved through computations or step-by-step algorithms, which resolves Hilbert's question of decidability.

When Hilbert first asked if mathematics was complete, consistent, and decidable, his hope was that the answer would be yes to all three. There is a Latin maxim *ignoramus et ignorabimus*. I know it kind of looks like a snooty schoolyard insult, but it actually translates to: "We do not know and we will not know." Hilbert thought this was a pessimistic view and that in a just world, there should be no *ignorabimus*—at least not in math. On his grave is an epitaph of his own defiantly contradictory motto, taken from his retirement speech in 1930: "We must know, and we will know."

Turing's work on the *Entscheidungsproblem* may have shaken the foundations of mathematics, but it also laid down the theoretical foundations for modern computers. People had built mechanical computing machines before Turing's time, but Turing was the first to think about algorithms and programs as mathematical concepts worth studying independently from the physical mechanisms used to house them and carry them out. This insight is what solidified his reputation as the father of computer science. The Turing machine is considered to be the basis for all modern computers. Even the most expensive, state-of-the-art supercomputers are no smarter than the Turing machine.

What started as a philosophical debate over the nature of reality and truth ended up inspiring the technology that would revolutionize civilization. These days, we carry around Turing machines in our pockets everywhere we go and rely on them for everything from emergency life support to ordering pizza.

It's amazing that the Turing machine isn't even Alan Turing's most famous accomplishment. Many people outside computer science probably recognize his name for his heroic efforts foiling enemy espionage during World War II. After the United Kingdom declared war on Germany in 1939, Turing went to work at Bletchley Park, a secret intelligence headquarters for codebreakers stashed away in a countryside manor. Turing was tasked with breaking German ciphers, and he became famous for breaking a particular code called the Enigma machine. This was a device that the Nazis used to encode phrases like "hello" into new strings like "jxrpm," which they could then broadcast because the only people who could decode the string of letters to reconstitute "hello" would be other Nazis with Enigma machines.

Turing's job was to break the code so that the Allies could listen in on German secrets. One of the Enigma machine's key flaws was that it would never encode a letter into itself, so the *h* in "hello" would never be encoded into another *h*. Turing and the rest of his team exploited this flaw to narrow down the possible Enigma settings, and then a large team of computers worked to hand-check the messages and search for fragments of common German words to build a key. Their successes gave the Allies a significant edge in the war effort, and some historians argue that the work of the mathematicians at Bletchley Park shortened the war by as much as two to four years.

Alan Turing's contributions can hardly be understated, but we must also recognize the contributions of the people who worked alongside him, including the many real-life computers who performed hundreds of hours of crucial mathematical calculations by hand. Women made up 75% of the workforce at Bletchley Park. One was Mavis Batey, a 19-year-old who studied German in university and found that her linguistic skills could be put to good use in the war effort. Prior to being recruited for Bletchley Park, she worked for the British secret service looking for coded spy messages printed in newspaper personal ads. She was so skilled at code-breaking that she was able to develop a new technique to decipher messages based solely on the information that two of the enemy operators each had a girlfriend named Rosa. Her colleague, Jane Fawcett, spurned her parents' pleas to attend her debutante ball. Rather than "come out" in society, she opted to work in Bletchley Park. She and several other women decoded a message that relayed the coordinates of the German battleship *Bismarck*, allowing the British Royal Navy to sink it with air strikes. She was behind the first major victory of the codebreakers.

History tends to credit big accomplishments to lone geniuses and heroes when the credit really ought to be shared by large teams of people who worked toward a common goal. This is especially problematic when those teams are predominantly made up of women or people of color. Even the supergroup's frontman, Alan Turing, didn't exactly receive the rich rewards or thanks he deserved from his government—because he was gay.

In 1952, Turing's house was burgled. When he reported the crime to the police, they discovered evidence that he was having a sexual relationship with a man. Homosexuality was a criminal offence in the United Kingdom at the time, so both Turing and his partner, Arnold Murray, were charged with gross indecency. That Turing was one of the nation's greatest mathematical minds, who had played a key role in saving his country from years of destruction and death, was little help in overcoming the British government's homophobia. In a single day, he went from being considered a war hero to a national security threat. Turing pled guilty and was given a choice between prison or probation with chemical castration. He chose probation. The chemical castration was essentially forced hormone therapy that reduced his sexual libido, made him impotent, and caused breast tissue to grow. His body became feminized against his will, and he was stripped of his security clearance and banned from ever working again for the British government, as well as ever entering the United States.

Two years after his trial, Turing was found dead in his home at age 41. An inquest determined that he had committed suicide by poisoning himself with cyanide. It wasn't until 1967 that homosexuality was decriminalized in Britain, and it took until 2009 for the British prime minister to issue an official apology to Turing on behalf of the government.

Queen Elizabeth II finally issued a royal pardon in 2013.

Alan Turing shouldn't have had to be a world-class genius to receive an apology and pardon. That he was human should have been enough.

In the last years before his conviction and death, Turing published a paper, "Computing Machinery and Intelligence," that addressed the topic of artificial intelligence (AI). This made him the first person to comment on AI aside from science fiction writers. Instead of tackling the question of whether machines can think, since Turing found the concept of "thinking" itself to be a hard thing to define, he replaced it with the question of whether machines could pass themselves off as humans. He proposed a hypothetical test called the imitation game, wherein a person would partake in a conversation with two other players, Player A and Player B. The subject would pose questions to both players, such as, "What was the last movie you watched?" or "What was your childhood like?" and each would respond. The twist is that either Player A or

Player B would be a machine. The subject's job is to determine which player is the human, based only on their responses to the questions. If the subject is unable to do this, the machine is said to be the winner.

The imitation game is now known as a Turing test, and you participate in one every time you have to identify all the squares that contain a traffic light or check the box that says, "I am not a robot," before accessing something on the web. These little quizzes, called CAPTCHAs, attempt to prevent bots and hackers from taking over websites. Most people don't realize the acronym stands for Completely Automated Public Turing test to tell Computers and Humans Apart. Interestingly, the origin of the Turing test was a party game where a player has to distinguish whether the hidden speaker they are talking to is a man or a woman. The reason the game is difficult, of course, is because the lines between genders are so arbitrary and porous.

Turing, writing about AI in 1950, had a prescient vision of an era when AI is more powerful and omnipresent than ever. Corporations can use AI to solve engineering problems, run election polls, or power self-driving cars. Individuals can hire AI assistants to write their emails and pitch work ideas, or even write academic journal articles. Students can use AI chatbots to write their essays, and artists can use AI programs to create incredible visual art. The lines between human-generated content and AI-generated content continue to blur. I can barely imagine what the state of AI might look like 50 years from now. At the very least, it seems inevitable that we will once again be reevaluating what it means to be real. How does a real human think? What does it mean for a being to have real consciousness?

Realness

Mathematicians and philosophers have spent thousands of years debating the nature of reality and objective truth. The story behind that body of discussion is a web of tragedies, technological marvels, and existential crises. We've discovered that there are things we simply can't know, and things we might feel are true but we can't prove, no matter how far our technology and intellect advance. Mathematicians will continue to debate the foundations of mathematics, and whether math is ultimately a creation of the mind, a game full of rules and symbols, or something else entirely.

The current debate over the true nature of gender is not too different from the debate over the true nature of mathematics. People like me blur the lines between genders, "passing" as a different gender from the one assigned to us at birth, with varying degrees of realness. When we leave our houses, it's like we are participating in Turing tests where everyone we encounter is trying to distinguish our "real" gender based only on our performance. But as the feminist philosopher Judith Butler writes, "gender is a performance."[4] Put differently, all the world's a stage, and the men and women are merely drag queens.

I unfortunately have no mathematically based resolution to offer here that will settle the question, What is "real"? I think Mama Ru put it best: "Real is what you feel."[5]

CHAPTER 10

Curtain Call

Math is drag. It is carefully calculated self-invention, and relies on courage and an open mind. To study it is to suspend your disbelief and accept a reality that's been curated with style, stamina, and wit. To perform it is to pursue beauty, and to innovate within it is to embrace creativity.

Contrary to popular belief, the goal of math isn't to get the right answer but rather to create a framework for understanding reality through patterns, abstractions, and metaphors. Drag does the same thing, but instead of lines and logic, we use glitter and glue. Both arts beckon you to break from the status quo and create an entirely new system, with new numbers, new geometries, and new rules. Math and drag alike are rooted in history and are at the same time futuristic. Fans and adepts of both must overcome intimidation to find enchantment.

Of course, math and drag are not entirely congruent. These days, math is typically appreciated as advantageous and useful—a benign or at least neutral force—while drag and all forms of gender nonconformity are highly controversial.

Queer people around the world are facing life-threatening systemic discrimination enforced by the highest powers there are: church and government. I'm grateful for the freedoms and privileges that allowed me to become the queen I am today, and I am painfully aware that many people do not share those freedoms—and that those freedoms are being taken away from queens and other queer people even as I write these words. I have the privilege of being able to wipe off the drag and pass as a cisgender gay man; I can even pursue whichever career path I aspire to without discrimination if I mask myself as straight. On the other hand, trans women are often unable to blend into the world of straight, cisgender people, and thus face the most discrimination, harassment, and violence. For Black trans women, the dangers are quadrupled; in addition to transphobia, homophobia, and misogyny, they also face persistent racism. Drag artists, trans people, and gender nonconforming people have always been the most visible and vulnerable part of the LGBTQ community.

Destroying the barriers and abuse we face will require collective action. That includes making our homes and classrooms safe places, where everyone feels included and free to be themselves. With that safety net in place, we can reach out to effect change in larger spheres, like businesses, universities, and municipalities. We will face resistance, but we must fight for our voices to be heard and our history to be shared. I still have this dream that leading with love is enough, and that using my drag to cultivate an enthusiasm for math can open minds to the reality that queer people are more than just a collection of stereotypes. I sometimes wonder, though, whether the powers of music, love, and visibility are enough to overcome our differences, with the climate around drag and trans people growing more hostile by the day. Can you change the world by changing your wig?

Prejudice is deeply entrenched even in places we think are neutral territory, like math. We like to tell ourselves that math doesn't discriminate, but it inherits from its creators all our social biases and prejudices, from the racist history of the IQ test to the political debate over the numberness of zero. One aspect of the work we should do to neutralize these biases is to provide a greater diversity of mathematical role models by uplifting the contributions of mathematicians from marginalized backgrounds. I'm working on this myself, as my own math education has been largely based on Eurocentric curriculums, which has made its way into my writing.

As crucially, if we are going to democratize math, we need to overhaul the way we teach it. We math lovers rhapsodize about how math is its own world, imbued with beauty and creativity, but the barrier to entry is standardized tests and timed competitions. We award scholarships and school placements to students who thrive in competitive environments and under time pressures, so students are left thinking that the only criteria for being good at math is to be able to think like a calculator. Perhaps instead of prioritizing correctness, memorization, and speed, we should be emphasizing the importance of creativity, collaboration, and deliberate, inquisitive thinking. I was lucky; my teachers supported me, respected me, and accepted me, even with all of my *bakla* antics, and they played a huge role in shaping the queen I've become. They fostered curiosity, kindness, skepticism, and creativity. Thanks are due to all the teachers going above and beyond to create and protect inclusive environments in their classrooms. These teachers need to be supported by systemic changes because currently they are contending with the enormous pressure of teaching to mandated curriculums and standardized tests, and working to protect their students' intellectual freedom and physical safety, all while trying to inspire more students than ever before.

For the next generation of problem solvers, creativity isn't optional, it's a matter of survival. The quest for knowledge, progress, and revolution operates in an unceasing cycle, whether in the ballroom or the classroom. There will always be new questions to ask, new metaphors to discover, and new realities on the horizon. The usefulness of math is certainly important but is only secondary to its main goal, which is the celebration of curiosity. That celebration is the driving force behind great math and great drag. Mathematicians and drag queens are artists who translate between the infinite and finite, imaginary and real, and random and fixed. Between those binaries, we find the thrill of discovery, the spark that dazzles the eye and ignites the soul.

ACKNOWLEDGMENTS

I couldn't have written this book without my editor, Tiffany Gasbarrini. I was hesitant about embarking on this journey until Tiffany convinced me that my story was worth writing. Thank you to everyone at Johns Hopkins University Press for taking a chance on this drag queen author.

I'd like to express my gratitude to all my teachers for shaping the person I've become, especially my math teachers for showing me the beauty of numbers, and my English teachers for showing me the might of the pen. Miss DeSousa, Mr. Wagner, Ms. Hesch, Mr. Macpherson, Mrs. DeVrieze: because of your support, I'll be a lifelong student.

I'd also like to thank my family. My mom is my hero, my little brother is my best friend, and my husband is my world. Appreciation is also due to my extended family for always being my biggest fans.

Finally, thank you to my dad, my very first math teacher.

THE "QUINTESSENTIALLY KYNE" PLAYLIST (ON RANDOM)

"I Surrender" by Celine Dion
"My Heart Will Go On" by Celine Dion
"Tell Him" (duet with Barbra Streisand) by Celine Dion
"It's All Coming Back to Me Now" by Celine Dion
"Greatest Love of All" by Whitney Houston
"I Will Always Love You" by Whitney Houston
"I Have Nothing" by Whitney Houston
"I Look to You" by Whitney Houston
"Try It on My Own" by Whitney Houston
"Didn't We Almost Have It All" by Whitney Houston
"All at Once" by Whitney Houston
"Total Eclipse of the Heart" by Bonnie Tyler

"Yesterday, When I Was Young" (1994 Remaster)
by Shirley Bassey
"I (Who Have Nothing)" by Shirley Bassey
"What Now My Love" by Shirley Bassey
"Climb Ev'ry Mountain" by Shirley Bassey
"This Is My Life (*La Vita*)" by Shirley Bassey
"Don't Cry Out Loud" by Melissa Manchester
"Faithfully" by Journey
"Open Arms" by Journey
"And I Am Telling You I'm Not Going"
(Original Broadway Cast 1982) by Jennifer Holliday
"I'd Do Anything for Love (But I Won't Do That)"
by Meat Loaf
"All Out of Love" by Air Supply
"Making Love Out of Nothing at All" by Air Supply
"When a Man Loves a Woman" by Michael Bolton
"Because You Loved Me" by Celine Dion
"Un-Break My Heart" by Toni Braxton
"I Turn to You" by Christina Aguilera
"Beautiful" by Christina Aguilera
"Reflection" Pop Version by Christina Aguilera
"Memory" by Barbra Streisand
"MacArthur Park Suite: MacArthur Park /
One of a Kind / Heaven Knows /
MacArthur Park Reprise" by Donna Summer
"Alone" by Heart
"If You Were a Woman (And I Was a Man)"
by Bonnie Tyler
"Loving You's a Dirty Job (But Somebody's
Gotta Do It)"
(with Todd Rundgren) by Bonnie Tyler
"Have You Ever Seen the Rain?" by Bonnie Tyler
"Straight from the Heart" by Bonnie Tyler
"The Winner Takes It All" by ABBA

"The Voice Within" by Christina Aguilera
"Without You" by Mariah Carey
"I Don't Wanna Cry" by Mariah Carey
"We Belong Together" by Mariah Carey
"I Know Him So Well" by Elaine Paige
and Barbara Dixon
"Maybe This Time" by Liza Minnelli
"Run to You" by Whitney Houston
"Saving All My Love for You" by Whitney Houston
"How Do I Live" Songbook Version by Trisha Yearwood
"I Am What I Am" by Douglas Hodge
"Don't Rain on My Parade" by Barbra Streisand
"All by Myself" by Celine Dion
"All the Man That I Need" by Whitney Houston
"Love Takes Time" by Mariah Carey
"My All" by Mariah Carey
"Listen to Your Heart" by Roxette
"That's the Way It Is" by Celine Dion
"Love Will Lead You Back" by Taylor Dayne
"Blue Bayou" by Linda Ronstadt
"Could It Be Magic" by Donna Summer
"Where Do Broken Hearts Go" by Whitney Houston
"The Power of Love" by Celine Dion
"Don't Stop Believin' " by Journey
"I Am What I Am" by Shirley Bassey
"Rolling in the Deep" by Adele
"Making Love (Out of Nothing at All)"
by Bonnie Tyler
"(Where Do I Begin) Love Story" 1994
Remaster by Shirley Bassey
"Hero" by Mariah Carey
"Could It Be Magic" by Barry Manilow
"One More Chance" by Air Supply
"Vision of Love" by Mariah Carey

"All I Wanna Do Is Make Love to You" by Heart
"These Dreams" by Heart
"What about Love?" by Heart
"Bohemian Rhapsody" Remastered 2011 by Queen
"Somebody to Love" by Queen
"We Are the World" by USA for Africa
"Original Sin" (Theme from *The Shadow*) Radio Edit by Taylor Dayne
"Fighter" by Christina Aguilera
"We Are the World" by Luciano Pavarotti & Friends
"You Raise Me Up" by Josh Groban

APPENDIX

CHAPTER 1, *page 2*

The number of blades of grass on Earth

If we assume there are 3 blades of grass per square centimeter, that's equal to $3 \times 100{,}000 \times 100{,}000 = 3 \times 10^{10}$ blades per square kilometer. Planet Earth has 148,940,000 square kilometers of land, and between 20% and 40% of that is covered in grass (depending on how you define grassland). If we are conservative and use the lower bound of 20%, then Earth has about 29,788,000 or 3×10^7 km^2 of grassland. Multiplying these together, the number of blades of grass is approximately $3 \times 10^{10} \times 3 \times 10^7 = 3 \times 3 \times 10^{(10+7)}$ which is approximately equal to 10^{18}.

CHAPTER 1, *page 7*

The probability of our monkey typing "LA ISLA BONITA"

The target phrase has 14 characters including spaces. The probability of *not* typing those 14 characters in a row is $1 - (1/29)^{14}$. If we try that 10^{20} times, we get:

$$\left(1 - \left(\frac{1}{29}\right)^{14}\right)^{(10^{20})} = 0.71 \text{ (approx)}$$

So there is a 71% chance of having never typed this 14-character string after 10^{20} attempts, and only a 29% chance of having typed it within that time period. After 10^{21} attempts, the probability of typing it rises from 29% to 97%.

$$\left(1-\left(\frac{1}{29}\right)^{14}\right)^{(10^{21})}=0.035 \text{ (approx)}$$

If each key press takes 1 second, then an "attempt" at this phrase takes 14 seconds. 10^{21} attempts would thus take 14×10^{21} seconds, or over 400 trillion years (443,633,230,000,000 to be exact).

CHAPTER 6, *page 145*
The Rule of 70

The exact formula for calculating the doubling time n (n is the number of time periods—years, months, hours, or whatever—used to express $R\%$) for a process growing at a rate of $R\%$ per time period is:

$$\left(1+\frac{R}{100}\right)^n=2$$

To solve for n, we will use an innovation called the logarithm, which essentially undoes exponentiation in the same way that division undoes multiplication, and subtraction undoes addition. Taking the natural logarithm, ln(), of both sides looks like this:

$$\ln\left(\left(1+\frac{R}{100}\right)^n\right)=\ln(2)$$

By the rules of logarithms, the exponent n can be taken out of the ln() function:

$$n\times\ln\left(1+\frac{R}{100}\right)=\ln(2)$$

And by isolating n, we get:

$$n=\frac{\ln(2)}{\ln\left(1+\frac{R}{100}\right)}$$

The numerator, $\ln(2)$, is exactly equal to an irrational number 0.69314. . . . But we can round this to 0.7, and the function $\ln(1+R/100)$ is approximately close to $R/100$ when R is between 0 and 100 (as most percentages are). So we can approximate $\ln(2)\ /\ \ln(1+R/100)$ with 0.70 / $(R/100)$, which simplifies to $70/R$. Thus the doubling time of a process which grows at $R\%$ per unit of time is approximately equal to $70/R$, an easier heuristic to remember!

NOTES

CHAPTER 1 *Infinite Possibilities*

1. *Paris Is Burning*, directed by Jennie Livingston (1990; Los Angeles, CA: Miramax, 2005), DVD.

CHAPTER 2 *Celebrity Numbers*

1. Girolamo Cardano, *Ars Magna or the Rules of Algebra*, translated by T. Richard Witmer (New York: Dover Publications, 1993).
2. Freeman Dyson, "Birds and Frogs," *Notices of the American Mathematical Society* 56: no. 2 (February 2009): 212–23.

CHAPTER 5 *The Average Queen*

1. "The Yentl Syndrome," *Washington Post*, July 27, 1991; Caroline Criado Perez, "Yentl Syndrome: A Deadly Data Bias against Women," *Longreads*, June 29, 2019, https://longreads.com/2019/06/21/yentl-syndrome-a-deadly-data-bias-against-women/.
2. Bernadine Healy, "The Yentl Syndrome," *New England Journal of Medicine* 325 (July 25, 1991): 274–76.
3. Jeffrey M. Jones, "US LGBT Identification Steady at 7.2%," Gallup.com website, February 22, 2023, https://news.gallup.com/poll/470708/lgbt-identification-steady.aspx.
4. Francis Galton, "Eugenics: Its Definition, Scope, and Aims," *American Journal of Sociology* 10 (July 1904), no. 1.
5. Francis Galton, "Africa for the Chinese," letter to the editor, *The Times*, June 5, 1873.

6. Theodore Roosevelt, letter to Edward Alsworth Ross, November 2, 1904, Theodore Roosevelt Papers, Library of Congress Manuscript Division, Theodore Roosevelt Digital Library, Dickinson State University, https://www.theodorerooseveltcenter.org/Research/Digital-Library/Record?libID=o189755.

CHAPTER 7 *Illegal Math*

1. Jaweed Kaleem, "How Drag Queen Story Hour Became a Battle over Gender, Sexuality and Kids," *Los Angeles Times*, February 22, 2023, https://www.latimes.com/world-nation/story/2023-02-22/drag-queen-story-hour.
2. Angele Latham, "Jackson Pride Event Continues to Spur First Amendment Debate," *Jackson Sun*, October 25, 2022, updated November 2, 2022, https://www.jacksonsun.com/story/news/2022/10/26/jackson-pride-event-first-amendment-debate-church-leaders-united-methodist/69581881007/.
3. " 'The Queen' Raided," *National Republican*, April 13, 1888, p. 1, https://chroniclingamerica.loc.gov/lccn/sn86053573/1888-04-13/ed-1/seq-1/.
4. Shane O'Neill, "The Stonewall You Know Is a Myth. And That's OK," *New York Times* video, see 4:15, May 31, 2019, https://www.nytimes.com/2019/05/31/us/first-brick-at-stonewall-lgbtq.html; Rebecca Onion, "Making Stonewall Matter," *Slate*, June 26, 2019, https://slate.com/human-interest/2019/06/stonewall-riots-controversy-primary-sources.html; Garance Franke-Ruta, "An Amazing 1969 Account of the Stonewall Uprising," *The Atlantic*, January 24, 2013, https://www.theatlantic.com/politics/archive/2013/01/an-amazing-1969-account-of-the-stonewall-uprising/272467/.
5. "Zazu Nova," Stonewall National Monument, National Park Service webpage, https://www.nps.gov/articles/000/zazu-nova.htm.
6. Anuj Misra, "Sanskrit Mathematics in the Language of Poetry," video lecture, Annual British Society for the History of Mathematics (BSHM) Gresham College Lectures, Gresham College, October 20, 2021, https://www.gresham.ac.uk/watch-now/sanskrit-mathematics.

CHAPTER 8 *Queer Geometry*

1. "Governor Ron DeSantis Hold [*sic*] a Press Conference in Bartow, Florida," The National Desk, Facebook, February 8, 2022, quote at 45:03, https://www.facebook.com/TND/videos/345369607469479.

2. Christina Pushaw, Twitter, @ChristinaPushaw, March 4, 2022, https://twitter.com/ChristinaPushaw/status/1499886619259777029.
3. Michelle Eder, "Views of Euclid's Parallel Postulate in Ancient Greece and in Medieval Islam," Rutgers University, Spring 2000, https://sites.math.rutgers.edu/~cherlin/History/Papers2000/eder.html.
4. "Quadrilateral of Omar Khayyam," Britannica Kids webpage, Encyclopaedia Britannica, 2003, https://kids.britannica.com/students/assembly/view/57049.
5. Péter Körtesi, "János Bolyai: The Founder of the Non-Euclidean Geometry," University of Miskolc, Hungary, presentation, 2015, quote on slide 14, http://old.pdf.upol.cz/fileadmin/user_upload/PdF/veda-vyzkum-zahr/2015/seminare/Janos_Bolyai__the_founder_of_the_Non-Euclidean.pdf.
6. János Bolyai, letter to Farkas Bolyai, November 3, 1823, transcribed and translated by Péter Körtesi, MacTutor Index, School of Mathematics and Statistics, St. Andrews University, February 2007, https://mathshistory.st-andrews.ac.uk/Extras/Bolyai_letter/.
7. Heartland Signal, Twitter, @Heartland Signal, March 6, 2023, https://twitter.com/HeartlandSignal/status/1632779718268157956.

CHAPTER 9 *Mathematical Realness*

1. *The Queen*, directed by Frank Simon (1968; New York, NY: Kino Lorber, 2020), Blu-ray.
2. "Realness," *Paris Is Burning*, directed by Jennie Livingston (1990; Los Angeles, CA: Miramax, 2005), https://www.youtube.com/watch?v=jHpt37S3wL8.
3. André Weil, quotations, MacTutor Index, School of Mathematics and Statistics, St. Andrews University, https://mathshistory.st-andrews.ac.uk/Biographies/Weil/quotations/.
4. Judith Butler, *Gender Trouble: Feminism and the Subversion of Identity* (Abingdon, UK: Routledge, 1990).
5. RuPaul, "The Realness," official music video, directed by Jason Whitmore, World of Wonder, May 15, 2016, https://www.youtube.com/watch?v=9jrUibTfAVI.

INDEX